# The SCIENCE of the
# OCEAN

NATURAL HISTORY MUSEUM

*The* SCIENCE *of the*
OCEAN

Penguin
Random
House

**DK LONDON**

**Senior Editor** Peter Frances
**Editors** Polly Boyd, Jemima Dunne,
Cathy Meeus, Annie Moss,
Steve Setford, Kate Taylor
**Managing Editor** Angeles Gavira Guerrero
**Production Editor** Kavita Varma
**Senior Production Controller** Meskerem Berhane
**Associate Publishing Director** Liz Wheeler
**Publishing Director** Jonathan Metcalf

**Senior Art Editor** Duncan Turner
**Designers** Francis Wong, Simon Murrell
**Photographer** Gary Ombler
**Illustrator** Phil Gamble
**Senior Jacket Designer** Akiko Kato
**Jacket Design Development Manager** Sophia MTT
**Managing Art Editor** Michael Duffy
**Art Director** Karen Self
**Design Director** Phil Ormerod

First published in Great Britain in 2020
This edition published in 2023 by
Dorling Kindersley Limited
DK, One Embassy Gardens, 8 Viaduct Gardens,
London, SW11 7BW

The authorised representative in the EEA is
Dorling Kindersley Verlag GmbH. Arnulfstr. 124,
80636 Munich, Germany

Copyright © 2020 Dorling Kindersley Limited
A Penguin Random House Company
10 9 8 7 6 5 4 3 2
003–316839–June/2023

A CIP catalogue record for this book is available from the British Library.
ISBN: 978-0-2416-3134-8

Printed and bound in China

**For the curious**
**www.dk.com**

MIX
Paper from
responsible sources
FSC™ C018179

This book was made with Forest Stewardship
Council™ certified paper – one small step in DK's
commitment to a sustainable future. For more
information go to www.dk.com/our-green-pledge

# Contributors

Jamie Ambrose is an author, editor, and Fulbright scholar with a special interest in natural history. Her books include DK's *Wildlife of the World* and *The Science of Animals*.

Dr Amy-Jane Beer is a biologist and naturalist. She began her career studying sea urchin development at the University of London before becoming a freelance science and nature writer and an advocate for wildlife and the natural world.

Derek Harvey is a naturalist with a particular interest in evolutionary biology, who studied Zoology at the University of Liverpool. His books include DK's *Science: The Definitive Visual Guide*, *The Natural History Book*, and *The Science of Animals*.

Dr Frances Dipper is a marine biologist and author and has been studying marine wildlife on the seashore and underwater, worldwide, for over 40 years. She has written numerous books for both children and adults. Her *Guide to the Oceans* won the Royal Society Aventis Prize for Junior Science Books in 2003.

Esther Ripley is a former managing editor who writes on a range of cultural subjects, including art and literature.

Dorrik Stow has a long career of research and publication on the oceans. He is Professor of Geoscience at the Institute of Geo-Energy Engineering Heriot-Watt University, UK, and Distinguished Professor at the China University of Geoscience, Wuhan, China.

# Consultant

Marine ecologist, author, and television presenter Maya Plass is Head of Communications at the Marine Biological Association, UK, and an honorary fellow of the British Naturalists' Association.

# Ⓝ NATURAL HISTORY MUSEUM

The Natural History Museum, London, looks after a world-class collection of over 80 million specimens spanning 4.6 billion years, from the formation of the Solar System to the present day. It is also a leading scientific research institution, with ground-breaking projects in more than 68 countries. Over 300 scientists work at the Museum, researching the valuable collections to better understand life on Earth. Every year more than five million visitors, of all ages and levels of interest, are welcomed through the Museum's doors.

Half-title page  Pale octopus (*Octopus pallidus*)
Title page  Giant devil rays (*Mobula mobular*)
Above  School of barracuda (*Sphyraena*) in the Red Sea
Contents page  Flower hat jellyfish (*Olindias formosus*)

# contents

# foreword

We manifest an intrinsic fascination with those things on the brink of our knowledge. There's a romance, a yearning, seemingly a need to wonder about things beyond our reach. Deep space, other planets, aliens... but, as is often cited, we may actually know more about these phenomena than about the inhabitants of our own oceans because 80 per cent of them are unmapped, unobserved, and unexplored for the so very simple reason that this environment is beyond your reach and mine. We are terrestrial, air-breathing mammals, and while some humans swim well, free dive, or use increasingly sophisticated scuba kit or robots to explore the oceanic unknown, the rest of us "land lubbers" just watch the waves.

Beneath those waves is another world, one that, thanks to science and technology, we are learning a lot more about, a lot more quickly. And, as this fabulous book shows, it is almost incomprehensibly beautiful and fantastic. From the smallest to the largest, the shallows to the deep, the fierce to the fearful, these organisms share our planet but live in a different dimension. How exciting!

So here is an opportunity to meet the neighbours, the "wet ones," the extraordinary diversity of miraculous life that has evolved in the oceans. And yet, remote as it can seem, the cultural aspects of our relationship with the sea reveal how we have always had a close connection to this charismatic, dangerous, and rewarding realm. But the tides have turned. Now we are the greater danger, and no drop of our seas is secure. Coral reefs are bleached. Plastic litters the greatest depths and fills the bellies of turtles and whales and chokes albatross chicks. The acidification of the water, pollution, and overfishing threaten the entire ocean ecosystem. There has never been a more important time to immerse ourselves in the wonder of the briny world and thus learn to love and protect it. Dive in, swim among stranger things, and then stand up for our oceans.

**CHRIS PACKHAM**
NATURALIST, BROADCASTER,
AUTHOR, AND PHOTOGRAPHER

SILKY, DUSKY, BLACK-TIP, AND GALAPAGOS SHARKS FEED WITH YELLOWFIN TUNA AND RAINBOW RUNNERS ON A SHOAL OF SCAD

# the marine world

**Almost as old as Earth itself**, oceans dominate the planet's surface. Life first evolved in the oceans, and today they are home to a vast diversity of species. By transporting huge amounts of energy, the oceans also help to power and modify Earth's climates.

# what are oceans?

Earth is a watery world, with 68 per cent of the surface covered by saltwater oceans. Each of the five major ocean basins (the Arctic, Atlantic, Indian, Pacific, and Southern) is a deep depression in Earth's surface. Connected to the oceans, and typically partly enclosed by land, are numerous smaller seas such as the Mediterranean and Bering seas. Oceans are partially separated by landmasses but are all interconnected.

The second largest ocean basin, the Atlantic spans around 20 per cent of Earth's surface

**Oceans from space**
Composite satellite images from space show clearly how oceans predominate over land. This view of Earth's western hemisphere shows the Atlantic Ocean.

**Ocean strength**
The oceans reveal their strength in the waves that crash onto shorelines around the world. Heading towards a shallow reef, this mountain-shaped wave off the south coast of New South Wales, Australia, will break as its base touches the seabed and frictional drag slows it down.

## EARTH'S PROPORTIONS

Oceans are immense, holding most of the water on Earth and having an average depth of about 3,700 m (12,000 ft) and a total volume of 1.34 billion cubic km (321 million cubic miles). Around 2 per cent of Earth's water is locked up as snow and ice, leaving 1 per cent as free-moving fresh water. The largest ocean is the Pacific, spanning 153 million square km (59 million square miles) and containing 49 per cent of Earth's salt water.

**KEY**

Salt water (68%)
Land (29%)
Fresh water (3%)

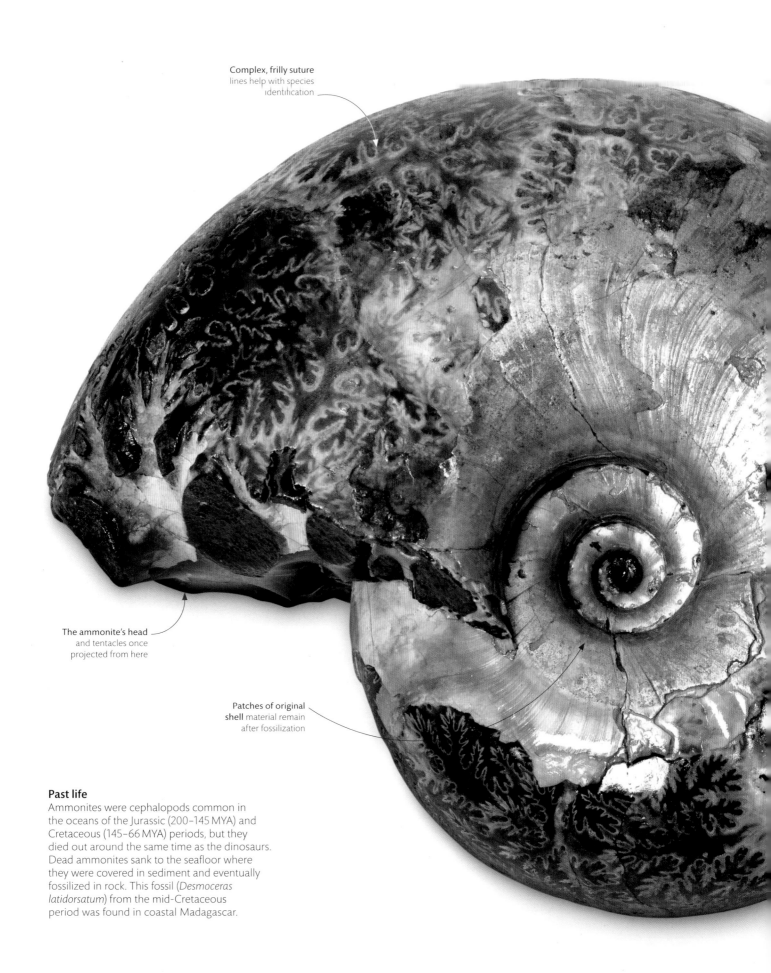

Complex, frilly suture lines help with species identification

The ammonite's head and tentacles once projected from here

Patches of original shell material remain after fossilization

## Past life

Ammonites were cephalopods common in the oceans of the Jurassic (200–145 MYA) and Cretaceous (145–66 MYA) periods, but they died out around the same time as the dinosaurs. Dead ammonites sank to the seafloor where they were covered in sediment and eventually fossilized in rock. This fossil (*Desmoceras latidorsatum*) from the mid-Cretaceous period was found in coastal Madagascar.

**Coral fossil**
Evidence that chilly Scotland was once part of a landmass near the equator, comes from warmth-loving fossil corals found there and dating to around 350 million years ago.

# ocean history

The familiar layout of Earth's continents and oceans is very different from the way it was in the prehistoric past. Today's oceans and their seas arose gradually as early landmasses split up and moved apart. Scientific evidence suggests that the original ocean formed from water vapour that escaped from the molten surface of the planet. Around 3.8 billion years ago, conditions cooled and this vapour started to condense and fall as rain.

## OCEANS OF THE PAST

Continents have come together and moved apart several times in Earth's history. Evidence for this comes from the "jigsaw" fit of continents such as South America and Africa, as well as palaeontology studies. Continents slowly change their positions through volcanic processes happening deep beneath the oceans. Today the Atlantic Ocean is getting bigger and the Pacific Ocean and Red Sea are getting smaller, but by less than 2 cm (³/₄ in) per year.

Cretaceous ocean

Cretaceous continent

LAURASIA

GONDWANA

**CRETACEOUS PERIOD C.130 MYA**

Foraminifera have thin shells called tests

**Ocean thermometers**
Analysing the chemical composition of sand-grain-sized fossil shells of microorganisms called foraminifera can help determine past sea-surface temperatures. Living foraminifera such as *Rosalina globularis*, above, are still found throughout the oceans.

**Stormy seas**
Oceans, atmosphere, and land all interact to create our climate and weather. Dramatic supercell thunderstorms over land or sea – here above Australia's Queensland coast – occur when wind conditions cause strong rotating updrafts. The frequency and intensity of extreme weather events is increasingly linked to human-instigated climate change.

# ocean climates

The world's oceans and their climates are inextricably linked through a complex web of interactions. The oceans absorb and store heat from the Sun, especially at the equator. From here, wind-driven surface ocean currents circulate water around each of the main ocean basins. In turn, the Sun's heat and Earth's rotation set the pattern of prevailing winds. Rainfall is highest in tropical areas, because the oceans are warmest here and so evaporation is high. The oceans slowly release heat, which is why the oceanic climate found around continental coastlines is less extreme than the climate far inland.

## CLIMATIC DISTURBANCES

Increased carbon dioxide in the atmosphere traps heat and causes ocean temperatures to rise as it absorbs extra heat. This can also lead to more extreme instances of the climatic disturbances known as El Niño and La Niña. The El Niño effect occurs where winds and currents in the equatorial Pacific are disrupted, leading to above-average sea-surface temperatures. The ONI records the difference between recent averages and the 30-year average temperature.

Values above 0.5 indicate warmer El Niño events

Values below -0.5 indicate cooler La Niña events

Oceanic Niño Index (ONI) °C

3.0 / 2.5 / 2.0 / 1.5 / 1.0 / 0.5 / 0.0 / -0.5 / -1.0 / -1.5 / -2.0 / -2.5 / -3.0

2008   2010   2012   2014   2016   2018   2020

Year

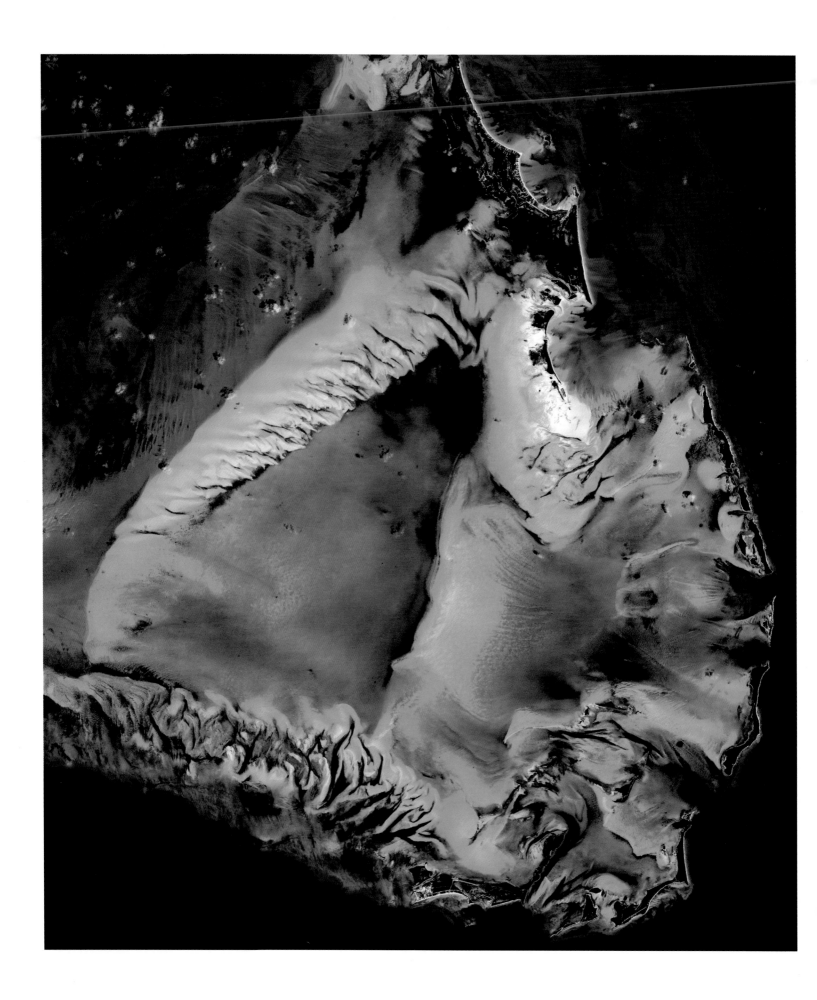

# ocean depths

Marine organisms inhabit a vast, three-dimensional space and can be found from the water surface down to the deepest depths. An organism near the surface will experience very different conditions to one in the deep ocean. Scientists divide the oceans into zones of decreasing sunlight and temperature, but pressure increases continuously from the surface to the deep. Surface layers are well-lit and have a good supply of nutrients and food. In deep zones, the sunlight fades and temperatures decrease, which limits life there.

**Deep ocean and sandy shallows**
In the Bahamas, the Berry Islands form a semicircle enclosing sheltered, sandy plateaus. This satellite image shows depth contours beneath the clear water. Deep ocean borders the land at the top, while Chub Cay at the bottom drops into the tongue of an ocean canyon.

Feeding tentacles
catch drifting plankton
and organic particles

## DEPTH ZONES
Marine life has adapted to the different ocean depths. Phytoplankton, the basis of almost all ocean food webs, only grow in the sunlit zone. In the twilight zone where light is limited, animals swim upwards at night to feed on plankton and then down for safety in the day. In the dark midnight zone animals produce bioluminescent light to hunt or communicate, while abyssal animals gather dead material or particles that drift down from above. Deep ocean trenches form the dark hadal zone.

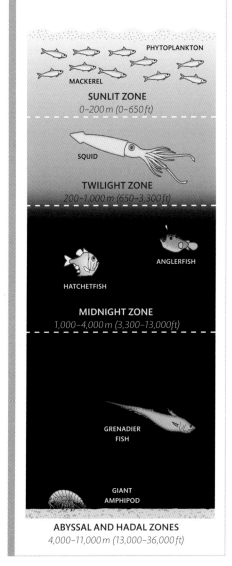

PHYTOPLANKTON

MACKEREL

**SUNLIT ZONE**
*0–200 m (0–650 ft)*

SQUID

**TWILIGHT ZONE**
*200–1,000 m (650–3,300 ft)*

ANGLERFISH

HATCHETFISH

**MIDNIGHT ZONE**
*1,000–4,000 m (3,300–13,000 ft)*

GRENADIER
FISH

GIANT
AMPHIPOD

**ABYSSAL AND HADAL ZONES**
*4,000–11,000 m (13,000–36,000 ft)*

**Cold-water coral**
Most corals live in shallow waters, but cold-water corals, such as *Lophelia pertusa*, form reefs in twilight and midnight zones. Without the nutrients provided by surface-level symbiotic algae, these corals grow very slowly.

# rocky coasts

**A rocky coast** provides a solid, dependable foundation for many plants and animals. Although there is often a risk of being swept away, a tidal pool or a high-rise cliff can also provide a refuge.

The branching **straps** of *Ramalina* form scattered clumps on the rocks

**Tufted lichen**
Grey *Ramalina siliquosa* and some other lichens grow as upright tufts. *Ramalina* usually lives at a higher level than yellow *Xanthoria* (see right) on European shores.

**Maritime lichen**
The sunburst lichen (*Xanthoria parietina*) of northwest Europe is typically found inland, but it can also tolerate the salty habitat of rocky shorelines. Its flat orange body, or thallus, may occasionally even be covered by the highest spring tides.

# surviving in the splash zone

Little life can thrive on wave-splashed rocks, which are too far from the reach of the tide for underwater organisms to survive and too hard for plants to root. But lichens have the resilience to grow here. In fact, they are so successful that in places they grow in prominent, colourful patches along the coastline. Key to their success is an intimate partnership: each kind of lichen consists of a nutrient-absorbing fungus that clings firmly to the rock and a pigmented alga that photosynthesizes to make food.

## INTIMATE PARTNERSHIP

The bulk of a lichen consists of the filaments of a fungus. Called hyphae, these filaments attach to the hard substrate and, as in other fungi, provide a large surface area for absorbing nutrients from organic food such as detritus and bird droppings. But more than 50 per cent of the lichen's food is made from atmospheric carbon dioxide by its photosynthetic algae, which share the food they produce with the fungus.

Algae and hyphae filaments disperse on wind to establish new lichen colonies

Algal cells

Fungal hyphae

Rootlike hyphae called rhizines attach the lichen to the rock

**SECTION THROUGH A LEAFY LICHEN**

The photosynthetic frond produces sugars, including mannitol, which is often exuded as a sweet powder on the frond's surface, thus inspiring this species' name

In the core of the stipe, filaments of conduction cells help to transport sugars around the kelp; seaweeds lack the transport vessels of true plants

The stipe, or stem, keeps the frond raised towards the light at high tide for photosynthesis

**Giant seaweed**

Sugar kelp (*Saccharina latissima*) of the northeast Atlantic grows particularly well when sheltered from extreme waves. Here, on the lower shore, its fronds can reach 4 m (13 ft) in length, which helps them extend further upwards to catch sunlight when lifted by the high tide.

The holdfast, unlike the roots of true plants, does not take up any nutrients, leaving the frond to absorb most of the material needed for photosynthesis

A crinkly edge
to the frond is
a characteristic
of the sugar kelp

**Holdfast communities**
Branching holdfasts can
provide a microhabitat
for diverse marine
invertebrates. Among
the holdfasts of kelp
growing on California's
Pacific coast live
scavenging brittle stars
and purple urchins that
graze on the seaweed.

Brittle stars cling
to the holdfast with
their sinuous arms

Purple urchins are
the dominant grazers
of California kelp

# anchoring to the seabed

Hard, rocky coastlines are impossible habitats for true plants to send down roots, but here – between the tides – conditions are perfect for the leafy algae popularly known as seaweeds. Instead of roots that penetrate, seaweeds have structures called holdfasts that cling to the substrate. Some holdfasts are like suckers, but others grow dense thickets of tendrils that anchor the seaweed to the seabed and shelter tiny invertebrates, while the alga's long, trailing fronds absorb light for photosynthesis.

## HOLDFASTS

A young seaweed develops into a frond, stipe (stem), and holdfast. The holdfast consists of branching, fingerlike projections called haptera, which envelop rocks and stones and become bushier in deeper water. It secretes mucopolysaccharide – a natural adhesive – as it grows larger, to support a bigger frond.

Growth
is mostly
horizontal

More
branches
to holdfast

Thick, bushy
growth

SHALLOWS        MID-DEPTHS        DEEP WATER

**Worn away by waves**
Over millennia the forceful churning action
of waves crashing against this coastline on
the Hawaiian island of O'ahu has unevenly
worn away the rock to create a network
of inlets and promontories.

# coastal erosion

The coastlines of every continent undergo constant change.
In some areas the land area expands, but in others the coastline
is worn away by erosion. Coastal erosion is determined by three
main interacting factors: marine energy (waves, storms, and
tides); the relative hardness of the rock; and tectonic activity
(earthquakes and uplift). The rate of coastal retreat can be more
than 100 m (330 ft) per century in areas where severe storms
pound a soft sand-and-clay shoreline. Hard granite cliffs may
remain stable for hundreds of years. Landslides, rockfalls,
chemical attack, wave impact, and relentless abrasion by sand and
gravel all help to break down rocks and boulders along the shore.
Storms, tidal currents, and rip-currents carry the sediment out to sea.

## HEADLANDS, ARCHES, AND SEA STACKS

Soft rock on a coastline is worn back by wave action into bays, while harder rocks
provide more resistance and form headlands, which are subject on three sides to
focused wave action. Eventually this creates notches and sea caves at the waterline.
The caves enlarge from either side of a headland, and the sea eventually breaks
through to form arches, which later collapse, leaving isolated sea stacks.

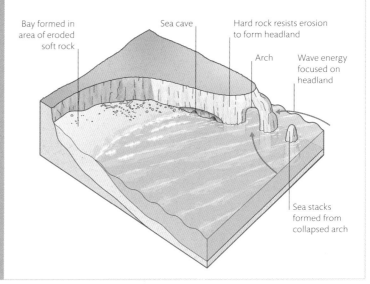

Bay formed in
area of eroded
soft rock

Sea cave

Hard rock resists erosion
to form headland

Arch

Wave energy
focused on
headland

Sea stacks
formed from
collapsed arch

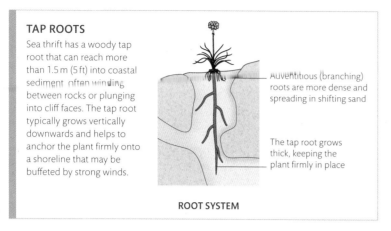

### TAP ROOTS

Sea thrift has a woody tap root that can reach more than 1.5 m (5 ft) into coastal sediment, often winding between rocks or plunging into cliff faces. The tap root typically grows vertically downwards and helps to anchor the plant firmly onto a shoreline that may be buffeted by strong winds.

Adventitious (branching) roots are more dense and spreading in shifting sand

The tap root grows thick, keeping the plant firmly in place

ROOT SYSTEM

### Tough plant

Sea thrift (*Armeria maritima*) thrives on windswept coasts around the northern hemisphere. By hugging close to the ground, its leaves are able to resist the elements, while its salt-tolerant tissues help the plant to grow in soil that may get occasionally inundated by extreme tides.

# growing above the tides

On the very edge of the ocean world, beyond the reach of normal tides, is a community of coastal species that belong to the land but live exposed to the sea's spray. Salt and wind are dehydrating: salt landing on foliage draws water from tissues by osmosis, while wind increases evaporation. The flowering plants that survive here must be adapted to resist these effects. Some, such as sea thrift, have thin, straplike leaves with a small surface area that reduces evaporation and a thick cuticle that traps water inside.

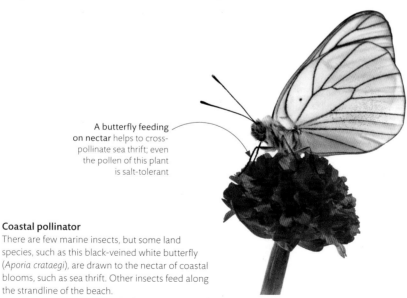

A butterfly feeding on nectar helps to cross-pollinate sea thrift; even the pollen of this plant is salt-tolerant

### Coastal pollinator

There are few marine insects, but some land species, such as this black-veined white butterfly (*Aporia crataegi*), are drawn to the nectar of coastal blooms, such as sea thrift. Other insects feed along the strandline of the beach.

TRIFT, ARMERIA VULGARIS WILLD.

Gen. Stab. Lit. Anst.

**Retracted tentacles**
reduce the animal's surface area, so there is less evaporation when it is uncovered by the retreating sea

**Dealing with exposure**
By pulling its tentacles into its body column, a beadlet anemone not only avoids predators but also retains more water while exposed to the air at low tide.

**Body column** contains muscle fibres that contract to pull the tentacles inside

# surviving low tide

The intertidal zone is the place where the ocean meets the land, but the living things that make it their permanent home come from the sea. Most of these organisms need to be underwater to breathe, feed, and reproduce. At low tide, those that can walk, such as crabs, can follow the water or shelter under rocks. Others that are fixed to the spot, including anemones, have no option but to wait for the waters to return.

**Tidal pool**
A pool on a rocky shore left by the ebbing tide is a safe haven for beadlet anemones (*Actinia equina*). Here, ocean animals can stay active, even when the surrounding rocks are left high and dry.

### INTERTIDAL ADAPTATIONS

These shore anemones illustrated in Ernst Haeckel's *Kunstformen der Natur* (1904) have variable intertidal adaptations. The daisy anemone (*Cereus pedunculatus*; top middle) lives partly buried and retracts completely when disturbed or exposed. Less-retractile species rely on a sheltered location to survive: the jewel anemone (*Corynactis viridis*; bottom left), drawn here with drooping tentacles, prefers rocky caves, while the plumose anemone (*Metridium senile*; bottom right), with feather-like tentacles, hangs limp from rocky overhangs at low tide. Such vulnerable species are more abundant where they are permanently submerged.

**HAECKEL'S SEASHORE ANEMONES**

**West Point, Prouts Neck (1905)**
US artist Winslow Homer painted his
favourite watercolour just after sunset,
capturing a column of seaspray leaping
above the rocks. It followed several days
of intense observation of the light and tides
across Saco Bay, close to the artist's studio
in Maine. The vivid reds and pinks, drenching
the sky and sea, and the bold composition
show leanings towards Expressionism.

**The Life Boat (1881)**
In the 1880s, Winslow Homer's art focused mainly on the struggle of fishing communities against the sea. This atmospheric pencil and wash study, which was the basis for a larger watercolour, depicts men in oilskins forging towards a ship in danger.

*the ocean in art*

# from coast to coast

In the second half of the 19th century, the US was a young country in search of an art culture that would reflect back the glory of a new nation. Artists sought to introduce audiences in the eastern states to the awe-inspiring mountains, plains, and seascapes to the west. The patriotic spirit captured in their work was reinforced by songs such as *America the Beautiful*, which celebrated a land that stretched "from sea to shining sea".

While Impressionist artists in Europe were rebelling against classical art forms, these were still preferred by many contemporaries in the US. The sublime landscapes of English-born American artist Thomas Cole, founder of the Hudson River School, an art movement that flourished in the mid-19th century, drew on Romanticism as well as realist and naturalist styles. The German-American painter Albert Bierstadt's grandiose peaks and plains, sketched during an expedition in the American West, were romantic constructs that sold idealized notions of the American Frontier to new settlers. He

also travelled to the West Coast, where he painted rugged rocks and crashing waves. Meanwhile, in the east, Fitz Henry Lane painted the coastlines of Maine and Massachusetts in Luminist style – a form of realist landscape painting characterized by precise brushwork and ethereal light.

Arguably, the pre-eminent American marine artist in the 19th century was Winslow Homer, known mainly for his oil paintings and watercolours. His early works depicted scenes of contemporary life in the US, but his art gained new depth after he spent 18 months on the British Tyne and Wear coast, where he painted the daily lives and hardships of local fishermen and women. He later retreated to a remote cottage and studio in Prouts Neck, on the East Coast in Maine, where he created magnificent seascapes that captured the energy and raw beauty of the ocean. Although he travelled and painted the iridescent seas off Bermuda and Florida, he always returned to Prouts Neck, where the ocean continued to provide a source of inspiration until his death in 1910.

> ❝ When you paint, try to put down exactly what you see. Whatever else you have to offer will come out anyway. ❞

WINSLOW HOMER, TO ARTIST FRIEND WALLACE GILCHRIST, *c.*1900

## THE MOLLUSCAN RADULA

A snail's "tongue" – correctly termed an odontophore – moves in and out of the mouth. In doing so, it rubs the radula over the substrate. The radula is covered with tiny teeth made from hard chitin (the same material found in invertebrate exoskeletons). As the teeth get worn down, the radula is regenerated, growing from a sac at the base of the odontophore.

Radula sac

Toothed radula is carried on the odontophore

Retractor muscle pulls odontophore into the mouth

Protractor muscle projects the odontophore

**RADULA STRUCTURE**

**Visible scratches** are left because the limpet's radula is harder than the surface rock

**Abrading limpet**
To dislodge the toughest algae, the radula of a limpet (*Patella vulgata*) has teeth reinforced with an iron compound. The material of these teeth is the hardest produced by any animal.

# rasping rocks

Sea-covered rocks are coated in a film of microorganisms, algae, and organic detritus, providing a source of food for many grazing animals. Dominant among them are molluscs, including a huge variety of herbivorous snails and limpets that rasp the surface as they crawl from place to place. Key to their success is their grazing equipment: a muscular "tongue" that scrapes the rocks with an abrasive ribbon called a radula, dislodging the nutritious material, which is then swallowed.

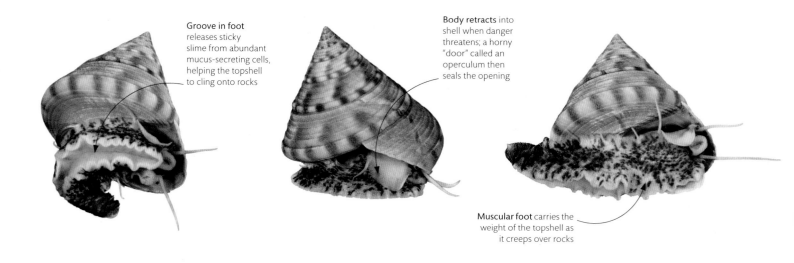

**Groove in foot** releases sticky slime from abundant mucus-secreting cells, helping the topshell to cling onto rocks

**Body retracts** into shell when danger threatens; a horny "door" called an operculum then seals the opening

**Muscular foot** carries the weight of the topshell as it creeps over rocks

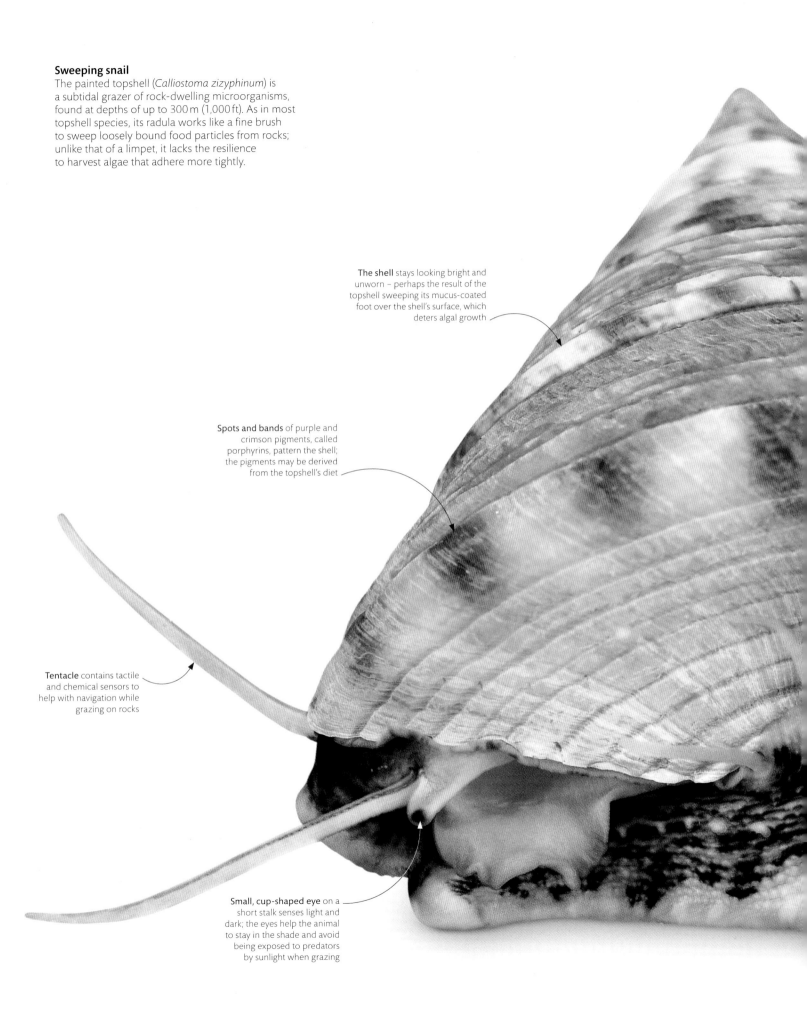

**Sweeping snail**
The painted topshell (*Calliostoma zizyphinum*) is
a subtidal grazer of rock-dwelling microorganisms,
found at depths of up to 300 m (1,000 ft). As in most
topshell species, its radula works like a fine brush
to sweep loosely bound food particles from rocks;
unlike that of a limpet, it lacks the resilience
to harvest algae that adhere more tightly.

**The shell** stays looking bright and
unworn – perhaps the result of the
topshell sweeping its mucus-coated
foot over the shell's surface, which
deters algal growth

**Spots and bands** of purple and
crimson pigments, called
porphyrins, pattern the shell;
the pigments may be derived
from the topshell's diet

**Tentacle** contains tactile
and chemical sensors to
help with navigation while
grazing on rocks

**Small, cup-shaped eye** on a
short stalk senses light and
dark; the eyes help the animal
to stay in the shade and avoid
being exposed to predators
by sunlight when grazing

**Attached by a thread**
California mussels (*Mytilus californianus*)
lack the limpet's clinging foot. Instead, they
tether themselves to rocks with a bundle
of protein fibres known as a byssus.

Fibres adhere to rock with
a kind of protein glue (a
different protein to that
forming the fibres)

# clinging on

Organisms that live on rocky shores exposed to battering waves are
constantly at risk of being washed away. Those that lack the agility to dodge
the breakers and find shelter in crevices must do what limpets do – cling on
tightly. Limpets are well equipped to survive the worst of the ocean's fury:
they have a cone-shaped shell that clamps down firmly, a muscular foot
that works like a suction cup, and sticky mucus that acts like glue.

**Intertidal life**
Common limpets (*Patella vulgata*) of
northwest Europe – seen here sharing their
rocky habitat with smaller periwinkles and
barnacles – hold fast at low tide during the
day. They emerge from their shells only when
submerged, or at night, to graze on algae.

### WAVES AND SHELL SHAPE

A limpet clamps down by contracting
muscles that pull the shell towards
the rock, reducing pressure under the
foot so that it functions as a sucker.
The foot also secretes sticky mucus
that helps to seal the attachment.
Where there is little wave action,
limpets grow flatter shells. Stronger
waves cause the limpet's muscles
to contract more in order to take
the extra strain. This affects the
shell-secreting tissues, producing a
taller and more conical shell shape.

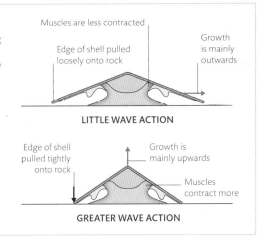

Muscles are less contracted

Edge of shell pulled
loosely onto rock

Growth
is mainly
outwards

**LITTLE WAVE ACTION**

Edge of shell
pulled tightly
onto rock

Growth is
mainly upwards

Muscles
contract more

**GREATER WAVE ACTION**

# combing for plankton

Acorn barnacles can be so abundant on a rocky shore that their small, volcano-shaped shells form a distinct white band. Each "volcano" opens at high tide so that the feathery limbs of the crustacean inside can reach up into the water. By quickly waving back and forth in a combing motion – or simply letting the backwash run through their "feathers" – the limbs catch nutritious detritus, algae, and the tiniest animals of the zooplankton.

## ZONES ON SHORELINES

Barnacles more tolerant of dehydration can live further up the shore, where low tide exposes them for long periods. In Europe, for example, *Chthamalus montagui* can survive above.*Semibalanus balanoides*, but it cannot thrive lower down due to competition from *S. balanoides* for space to feed. Similar barnacle zonation occurs around the world.

Smaller *Chthamalus montagui* predominate on upper shore

**HIGH TIDE**

Larger *Semibalanus balanoides* predominate on middle shore

Predation and competition from other animals prevents barnacles colonizing lower shore

**LOW TIDE**

## Casting the net

Magnified many times, the thoracic limbs, or cirri, of a grey acorn barnacle (*Chthamalus fragilis*), from eastern North America, are seen to carry long hairs, called setae. The setae overlap to create a trap that catches particles as small as two-thousandths of a millimetre in size.

**A diamond-shaped aperture**, here closed at low tide, is characteristic of the northern acorn barnacle (*Semibalanus balanoides*) of Europe and North America

## Exposed and closed

Barnacles cannot feed at low tide, so they withdraw their limbs, then close their shells using moveable plates that seal the entrance. Like this, they are protected from desiccation until the tide returns and they become active again.

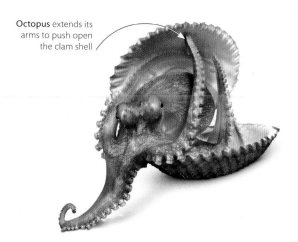

**Octopus** extends its arms to push open the clam shell

**Flexible arms** coil around the body so the octopus can fit into the shell

**Suckers** grip the shell as the arms pull it closed

**Octopus in a shell**
Hermit crabs are not the only animals to use shells for protection. Along tropical shores, the coconut octopus (*Amphioctopus marginatus*) sets up home in a variety of objects, including coconut shells and, as here, the hinged shells of clams.

**The crab's borrowed shell** is often colonized by other animals; this whelk shell is covered in snail fur (*Hydractinia echinata*), a kind of colonial hydrozoan

**Two larger pairs of legs** are used for walking; two smaller pairs, hidden from view, are used to grip the inside of the shell

# lodging in a shell

Only the top of a hermit crab's head is shielded by a hard carapace; the rear half of its body is soft and vulnerable. For this reason, hermit crabs spend their lives lodged for protection inside a coiled snail's shell. They use their strong limbs to drag this mobile sanctuary with them wherever they go. When a crab grows too big for its shell, it seeks an upgrade – carefully testing the new shell's size and weight before making the switch.

## CRAB ANATOMY COMPARED

Most crabs have a short, flaplike abdomen that folds beneath the hard carapace of their body and is all but concealed. But in hermit crabs, the long, soft abdomen twists to one side, which enables it to fit more easily into a coiled snail shell. Tiny, hook-shaped structures called uropods at the abdomen's tip and small legs along its sides grip the inside of the shell to hold the crab in place.

**KEY**

- Carapace
- Abdomen
- Walking leg
- Claw/pincer
- Telson
- Uropod

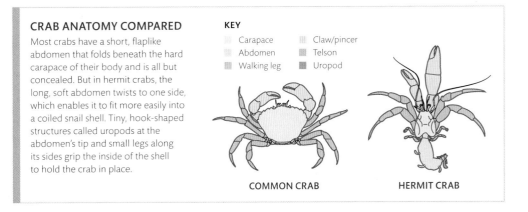

COMMON CRAB

HERMIT CRAB

### Shells for size

In their seashore habitat, juvenile European common hermit crabs (*Pagurus bernhardus*) make use of the most common empty shells found there, including those of periwinkles and dogwhelks. In deeper water, older crabs use the shells of larger snail species.

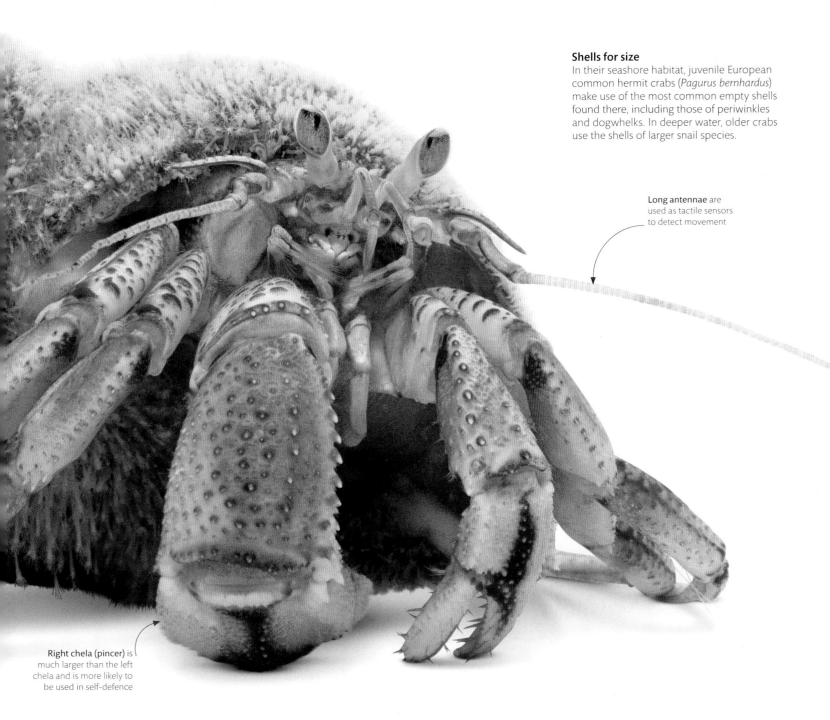

**Long antennae** are used as tactile sensors to detect movement

**Right chela (pincer)** is much larger than the left chela and is more likely to be used in self-defence

**Breaking wave**
This image shows the moment when a wave formed in the open ocean approaches the shore and the rolling "tube" of water collapses to form a breaking wave.

# waves

A wave is a disturbance in the surface of the sea produced by the transfer of energy from wind to water. Wind-generated waves range in size from a barely ruffled surface and snaking ripples to huge rogue waves reaching a height of up to 10 m (33 ft) that can snap a giant tanker in half. Major storm waves generated in the Southern Ocean around Antarctica can travel for nearly two weeks across the Pacific before breaking on the remote shores of Alaska. Their size is diminished, but their distinctive pattern remains the same as when first formed. The amount of energy stored in waves is enormous – a single storm wave can exert up to three tonnes per square metre as it breaks.

## WAVE FORMATION

The three stages in wave development are known as sea, swell, and surf. The state of random choppiness building to larger irregular waves is known simply as sea. As these waves leave the region in which they were generated, wave interaction slowly develops a distinctive pattern of waves known as swell. Eventually, as a wave train enters shallow water, it interacts with the seafloor. The movement slows, and the distance between the waves reduces, resulting in a relative increase in wave height. Eventually, the ratio of wave height to wavelength reduces to the point where the wave topples forwards and breaks as surf.

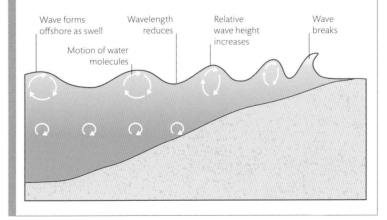

Wave forms offshore as swell

Wavelength reduces

Relative wave height increases

Wave breaks

Motion of water molecules

**Flexible hunter**

The small epaulette shark (*Hemiscyllium ocellatum*) has several adaptations that allow it to hunt in very shallow intertidal pools, where it targets worms, crabs, shrimps, and small fish. The shark is 70–90 cm (28–35 in) long, slender, and flexible, and thus able to manoeuvre easily in small spaces using its sturdy fins. Its flattened underside and blotchy coloration allow it to lie undetected until prey ventures within range.

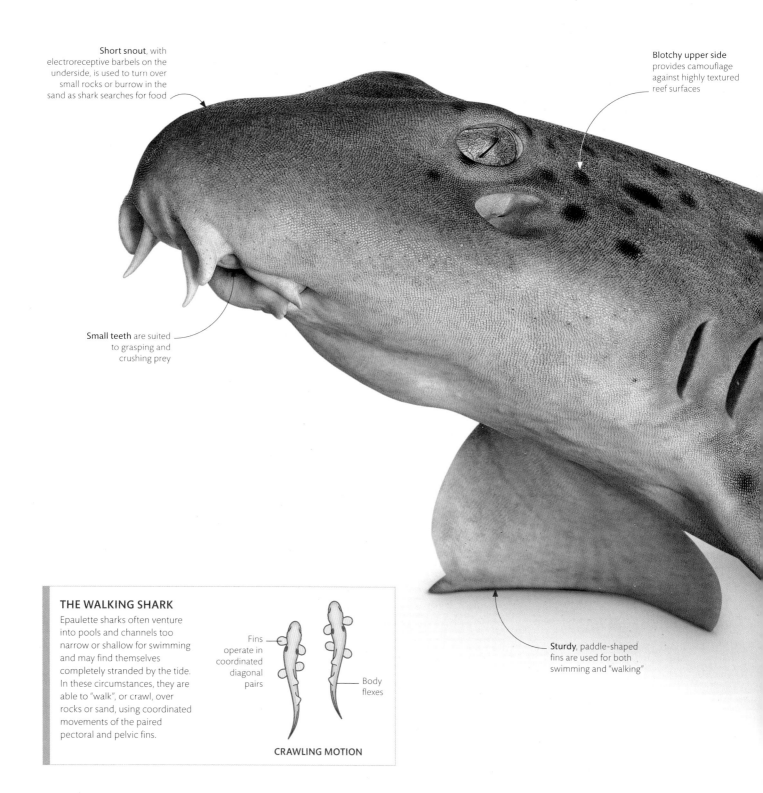

**Short snout**, with electroreceptive barbels on the underside, is used to turn over small rocks or burrow in the sand as shark searches for food

**Blotchy upper side** provides camouflage against highly textured reef surfaces

**Small teeth** are suited to grasping and crushing prey

**Sturdy**, paddle-shaped fins are used for both swimming and "walking"

**THE WALKING SHARK**

Epaulette sharks often venture into pools and channels too narrow or shallow for swimming and may find themselves completely stranded by the tide. In these circumstances, they are able to "walk", or crawl, over rocks or sand, using coordinated movements of the paired pectoral and pelvic fins.

Fins operate in coordinated diagonal pairs

Body flexes

**CRAWLING MOTION**

# fish on land

The intertidal zone is a challenging habitat, with extreme fluctuations in temperature, salinity, and oxygen levels. But it also offers rich feeding opportunities and safety from large marine predators. Fish that live in these waters include the epaulette shark, which can last a few hours out of water, and mudskippers, which can spend up to 18 hours a day on land. To cope with oxygen-depleted waters at low tide, they shut down non-essential metabolic functions, reduce their heart rate and blood pressure, and prioritize the supply of oxygen to the brain.

**Eyes** can retract into water-filled skin folds to prevent them drying out on land

**Fish out of water**
Found on mudflats, in estuaries, and in swamps, mudskippers (here *Periophthalmus barbarus*) are the only fish to walk, eat, and court on land. They survive by carrying water in gill chambers.

**Prominent** shoulder markings (or "epaulettes") resemble huge eyes and may deter predators

**Flexible** spinal column allows movement in tight spaces

**Flattened underside** allows the shark to press itself to the seabed without casting a tell-tale shadow and provides stability on land

**Tapering** at head and tail permits fast open-water swimming

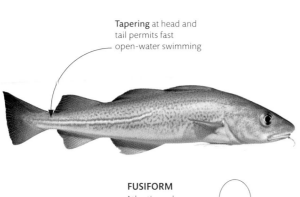

**FUSIFORM**
Atlantic cod
*Gadus morhua*

**Lateral flattening** gives quick bursts of speed, sharp turning, and the ability to fit through narrow spaces

**COMPRESSIFORM**
Emperor angelfish
*Pomacanthus imperator*

**Flattened shape** enables fish to live on the seabed; it flaps its fins up and down like a bird for swimming

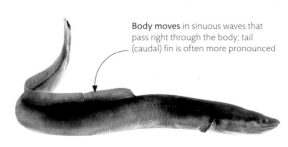

**DEPRESSIFORM**
Pacific leopard flounder
*Bothus leopardinus*

**Spherical-bodied** fish are slow swimmers that rely on camouflage and adaptations to avoid predators and capture prey on the seabed

**GLOBIFORM**
Warty frogfish
*Antennarius maculatus*

**Body moves** in sinuous waves that pass right through the body; tail (caudal) fin is often more pronounced

**ANGUILLIFORM**
European eel
*Anguilla anguilla*

**Elongated beak** and arrowlike body enables fish to swim fast over short distances

**SAGITTIFORM**
Longnose gar
*Lepisosteus osseus*

**Thin, threadlike** body slithers through water like a snake

**FILIFORM**
Hair-tail blenny
*Xiphasia setifer*

**Ribbon-shape** enables fish to hide in rock crevices, but it is a slow swimmer

**TAENIFORM**
Crescent gunnel
*Pholis laeta*

**Body** can be up to 1.5 m (5 ft) long, but weighs only 170–200 g (6–7 oz)

**Females and immature males** have a birdlike beak that curves outwards and cannot be fully closed

**Dorsal fin** runs full length of the body

# fish shapes

The shape of a typical fish reflects the fact that it moves through water – a medium more dense than air – and is affected less by gravity than land animals. A smooth streamlined body, with a rounded cross section and pointed front end, is better for cutting through water, while fins (rather than weight-supporting limbs) control position or help generate thrust. But many fish have departed from this ideal shape to exploit specialized life styles.

### Extraordinary lengths
The pale snipe eel, or deepsea duck (*Nemichthys scolopaceus*), has around 750 bones in its spine – more than any other animal on Earth. It is found in oceans the world over at depths of 90–1,800 m (300–6,000 ft). Observations suggest that it orientates itself vertically and feeds entirely on mid-water pelagic crustaceans.

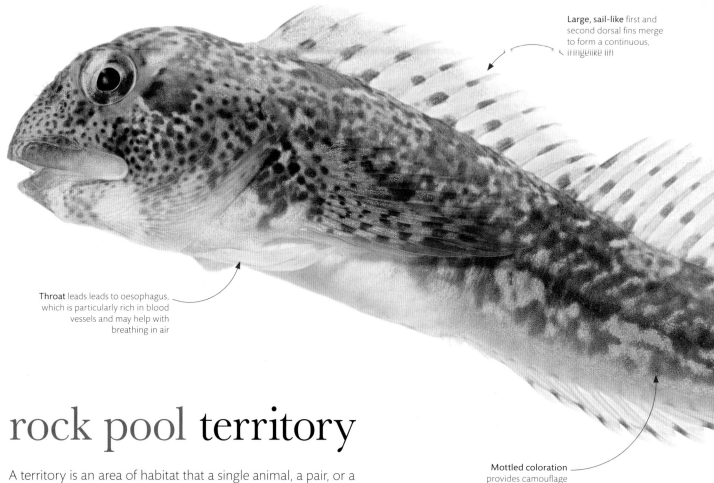

**Large, sail-like** first and second dorsal fins merge to form a continuous, fringelike fin

**Throat** leads leads to oesophagus, which is particularly rich in blood vessels and may help with breathing in air

**Mottled coloration** provides camouflage and becomes much darker in brooding male

# rock pool territory

A territory is an area of habitat that a single animal, a pair, or a group defends from others of the same species or other species, earning them exclusive access to food, mates, and hiding places. For the shanny (*Lipophrys pholis*), the advantages of a rock pool territory cut off from others at low tide balance the challenges of life in the intertidal zone (see p.45). Several female shannies may deposit eggs in a pool guarded by one male. At high tide, the male will attack almost any animal venturing near.

**Large pectoral fins**, with stout bases, provide leverage for moving on land

**Shanny on land**
If stranded by the tide, the shanny hides in crevices or moves over rocks. Its tail sweeps back and forth with enough traction on the ground to help push the fish forwards.

**Caudal fin** provides some propulsion during swimming when swept from side to side

### TYPES OF FINS

All bony fish share a similar arrangement of dorsal, anal, and caudal fins along the midline and paired pelvic and pectoral fins. However, some fins are modified according to the species' mode of living. For example, the fins of the shanny enable it to swim quickly and strongly against tidal currents and to crawl over land.

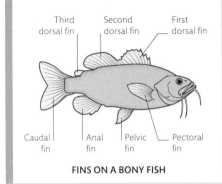

Third dorsal fin — Second dorsal fin — First dorsal fin

Caudal fin — Anal fin — Pelvic fin — Pectoral fin

**FINS ON A BONY FISH**

### Living with limits

The shanny's ability to exchange respiratory gases in air as well as water is a vital adaptation to the oxygen-depleted environment of a rock pool. At low tide, when feeding opportunities are limited, the fish adopts a relatively low-energy lifestyle, lurking hidden until high tide or until roused to territorial defence.

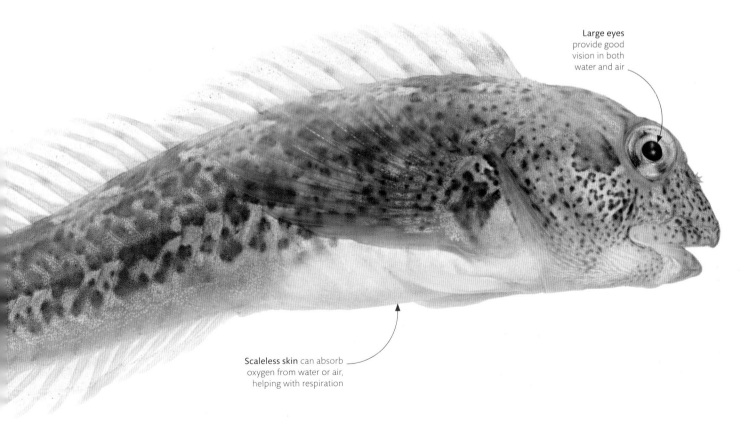

**Large eyes** provide good vision in both water and air

**Scaleless skin** can absorb oxygen from water or air, helping with respiration

# cold-blooded diving

A lizard that relies on the Sun's rays to stay warm and active is an unlikely cold-water diver – but along the rocky coastlines of the Galápagos Islands in the eastern Pacific Ocean, the marine iguana (*Amblyrhynchus cristatus*) does just that. On this volcanic archipelago, where competition for land-sourced food is intense, the marine iguana has evolved to eat seaweed. Females and juveniles usually graze between the tides. Large males dive deeper, enduring the chilly Humboldt Current, which comes straight up from Antarctica and streams around these islands.

**Blunt face** allows front teeth to get close to rock when grazing seaweed

**Specialized teeth**
The iguana's jaws are lined with three-pronged teeth, which spear the seaweed fronds that the lizard rips from the rock.

**The skin's covering of beadlike scales** protects the body from physical injury and reduces water loss by evaporation

**Dark skin patches** absorb the Sun's radiation well, warming the blood

## SNEEZING SALT

A diet of seaweed quickly loads the body with too much salt, which would disturb the body's salt/water balance, damaging cells. Sea reptiles and seabirds possess glands that actively pump salt from the blood into salt-secreting glands, from where it leaves the body. The marine iguana, like seabirds, has paired nasal salt glands: the excess salt mixes with mucus and is periodically forced out by sneezing through the nostrils.

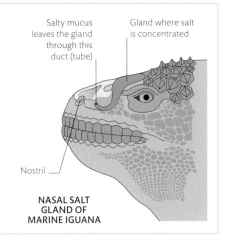

Salty mucus leaves the gland through this duct (tube)

Gland where salt is concentrated

Nostril

**NASAL SALT GLAND OF MARINE IGUANA**

Conical spines on **the head** develop in adult, sexually mature iguanas

A white crust of salt, expelled from glands in the nostrils, often covers the top of the head

### Cold-blooded diver

Like all reptiles, the marine iguana is ectothermic – meaning that it depends upon its surroundings for its warmth. It can last for just over an hour in the cold sea around the Galápagos. Then, before it chills too much to function, the iguana must return to land to bask in the Sun to recharge with solar heat.

**A Fishing Boat at Sea (1888)**
The emotional energy captured in the most simple brushwork in Vincent van Gogh's work was an inspiration to the Fauves.

**Port-Miou (1907)**
Georges Braque's headlong plunge into Fauvism is signalled in his passionate colouring of the high ridges and trees fringing the long inlet of Port-Miou on the coast near Marseille, France. There is a suggestion of Cubism in the diamonds on his cobalt-blue sea.

*the ocean in art*

# seas of colour

When the first Fauve paintings went on show in 1905 at the Salon d'Automne in Paris, France, a critic compared them to the "naive games of a child playing with a paint box". Here were landscapes, seascapes, portraits, and nudes that dispensed with all notions of representation and fidelity to nature. Colour was redefined: flattened shapes of saturated hues were elements of the painting in their own right and vehicles to express the artists' emotions.

The Fauves drew their inspiration from the late 19th-century Post-impressionists, such as Vincent van Gogh, who brought an ecstasy and anguish to colour and brushwork that transformed his subjects. In 1881, Paul Gauguin urged fellow artist Paul Sérusier to use not just green for a tree, or blue for a shadow, but the most *beautiful* green and the *bluest* blue.

Fauve theories, which would underpin Expressionist, Cubist, and modern art to the present day, emerged while Henri Matisse and André Derain spent the summer in the fishing port of Collioure, France, in 1905. They experimented with line and colour to capture the enhancing effects of Mediterranean light. Similarly, Georges Braques produced his first Fauvist work in L'Estaque, near Marseille, in 1907. All three artists flooded their canvasses of ports and coasts with bold colours and simplified forms that expressed their subjective response and intensity of feeling rather than objective reality.

    **We were always intoxicated with colour, with words that speak of colour, and with the sun that makes colours live.** 🙳

ANDRÉ DERAIN, 1880–1954

**Rictal rosette** is a fleshy flap around the base of the bill's gape; it gets bigger and turns yellow-orange in breeding birds

**The bill's horny outer sheath** – the ramphotheca – grows thicker and more pigmented in breeding puffins, making the bill bigger and more colourful; this helps to reinforce the bond between monogamous pairs

**Counter-shaded body plumage** – black above and white below – may help to disguise the bird against ocean water when fishing at sea, protecting it from both aerial and underwater predators

**Broad bill** enables each parent to hold many sand eels – crosswise – at once

**Clifftop chick**
Like puffins, the closely related razorbill (*Alca torda*) divides responsibility when raising its brood: both parents feed the chick with sand eels caught by diving.

### EGGS ON LEDGES

Eggs are vulnerable in a cliff-face colony. Guillemots – relatives of puffins – lay spotted, pear-shaped eggs on exposed ledges, contrasting with whiter, rounder, typically spot-free puffin eggs hidden in crevices. Distinct spotted patterns may help guillemot parents to recognize their own eggs. The eggs' less-rounded shape lets them sit flatter and more securely on the nest and spreads the stress of knocks from jostling birds over more of the egg surface, reducing the risk of breakage.

**GUILLEMOT EGGS**

# cliff nesting

Many birds depend entirely upon the ocean for food, but all must return to land to breed and lay their hard-shelled eggs. Flight makes a range of coastal habitats easily accessible, including steep cliffs and rockfaces. On precarious ledges and among rocks – safe from non-flying predators – puffins, guillemots, gannets, and other birds gather to pair up and raise the next generation.

**A dark fleshy "horn" projects** above each eye in sexually mature adults, unlike in puffins of the North Atlantic

**Colourful colonist**
In summer, the horned puffin (*Fratercula corniculata*) sports its breeding colours: its winter grey face becomes an immaculate white mask, and its bill turns yellow and bright red. Around north Pacific coastlines, horned puffins gather in colonies to nest on rocky cliffs. The densest colonies occur where there are plenty of crevices and cavities to shelter the eggs and chicks.

# northern gannet

With a wingspan of up to 1.8 m (6 ft), the northern gannet (*Morus bassanus*) is the north Atlantic's largest seabird. A strong flier, this bird spends most of its life at sea, regularly making 540-km (336-mile) foraging expeditions; tagged birds have been recorded making even longer journeys. It soars over northwest Europe's coastal waters year round, but its range extends from the southern Arctic to the Gulf of Mexico on a seasonal basis.

Oily fish and squid comprise most of a gannet's diet – herring, mackerel, and sprats are its preferred prey – but, as opportunists, these birds also steal any type of fish from nets and other seabirds. They often follow fishing boats far out to sea in the hope of feasting on discarded bycatch. More usually, the northern gannet stays closer to shore, flying above the water at heights of 10–40 m (33–130 ft) at an average speed of around 15 kph (10 mph). It searches the surface for schools of fish, identifying food sources by tracking dolphins or larger predatory fish. Some gannets hunt alone, but they typically feed in flocks of up to a thousand. When they spot their prey, hundreds plummet head-first into the sea, often from a considerable height, to grab a meal.

Pairs of gannets mate for life, reinforcing bonds through behaviours such as "mutual fencing" (knocking together of bills), and they return each year to nest in large, raucous colonies, known as gannetries. Although there are a few sites around North America, most northern gannets breed in 32 gannetries located between Brittany, France, and Norway that contain more than 70,000 breeding pairs. The female lays a single egg in late spring in a mounded nest of seaweed, feathers, and plant matter. Parents take turns incubating the egg by covering it with their feet, and they vigorously defend nests and young by stabbing intruders with their bills.

### Diving for food

Gannets hunt by plunging into the sea beak-first, with their wings folded back, at speeds of up to 86 kph (53 mph). Most plunge dives are relatively shallow, but they can be as much as 22 m (72 ft) deep.

### Rocky roost

Gannetries are usually located on rocky coastlines, protected by the rugged terrain, like this one on the Shetland Isles. Although adult gannets have few natural predators, their eggs and chicks may be stolen by gulls, ravens, foxes, or weasels.

# hopping on rocks

For most seabirds that can fly, the journey between water and land typically presents few problems. But for flightless penguins, a rocky shoreline battered by waves is a constant challenge. Rockhopper penguins must brave the choppy waters as they dive for krill, and then – using little more than their clawed, webbed feet and willpower – scale the rocks on their return.

**Vestigial hind toe** does not reach the ground

**Sole-walkers**
Penguins have plantigrade feet, which means that they walk on the soles of their feet, giving better traction. Other birds are digitigrade – they walk on their toes.

**Island penguin**
Southern rockhopper penguins (*Eudyptes chrysocome*), of the Falklands and other subantarctic islands, thrive on coastlines jagged with solidified lava, where they clear steep gullies in single jumps. The habitat, although harsh and craggy, provides shelter for the birds' nests and pools of fresh drinking water that collect between the rocks.

**Short, stout, bill**, which is used for grabbing krill when diving, may also be used like a grappling hook when climbing steep slopes

**Three forward-pointing toes** have thick, strong claws that can grip rock; the toes are connected by webbing for surface paddling

**Short, thick legs** are powered by strong thigh muscles that help the bird to hop from rock to rock

## PENGUIN POSTURE

A streamlined body, with legs set far back, enables a penguin to cut through water like a torpedo. But adaptations that perfect swimming and diving leave a penguin ungainly on land. Compared with a gull, the legs of a penguin are lower down on the body, forcing the bird to stand more upright and giving it a waddling gait.

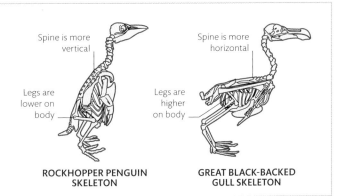

Spine is more vertical

Legs are lower on body

**ROCKHOPPER PENGUIN SKELETON**

Spine is more horizontal

Legs are higher on body

**GREAT BLACK-BACKED GULL SKELETON**

**Stiff, flat, wings** are held outwards for balance when jumping; they are used as paddles for propulsion when the bird dives

# sandy beaches

**The shifting surface** of a sandy beach
is a difficult place to put down roots or dig
a burrow, but it also presents opportunities.
Flotsam arriving on a strandline attracts
scavengers, and gently sloping beaches allow
large, air-breathing animals to come ashore.

**Making dunes**
Wind-blown sand gathers around plants and other obstacles to create dunes. With their deep network of roots, marram grasses stabilize dunes and enable them to increase in size.

**Lines of new plants** are produced from rhizomes that grow horizontally through the sand

# growing in wind-blown sand

A sandy seashore exposed to gales can be a harsh environment for a plant: the salty wind dries tender shoots, and the sand it carries scours leaves and buries new growth. In these tough conditions, marram grasses not only survive – they thrive. They have wind-resistant leaves and a regrowth system that continually pushes the plants upwards from beneath rising piles of sand.

**Upper epidermis** (surface layer of cells) is exposed to the wind, but it has a thick, waxy cuticle that reduces water loss from evaporation

### STIMULATED BY BURIAL

Most short plants quickly get covered by shifting sand, but marram grass continually sends up vertical rhizomes (underground stems) from its rootstock and sprouts new green leaves above the sand's surface. As more and more sand accumulates, the plant develops layer upon layer of additional roots higher in the sediment, which binds the sand together as the root fibres get thicker.

New leaves

Rootstock

Rhizomes

Second-year root layer

First-year root layer

**GROWTH PATTERN OF AMMOPHILA ARENARIA AFTER BURIAL IN SAND**

### Wind tolerance

A cross section through common marram grass (*Ammophila arenaria*) shows how the leaf blade is rolled inwards, sheltering the stomata – minute pores that exchange gases for respiration and photosynthesis. The curled shape screens the stomata from the drying effects of the wind, while the central channel keeps moist air trapped close to more vulnerable, porous tissue layers.

**Lower epidermis** is rolled inwards and carries hairs to reduce air movement

**Stomata (pores)** for gaseous exchange are sunk at the base of pits, where they are less likely to lose water in the drying wind

**Green mesophyll tissue** contains the pigment chlorophyll, which absorbs light energy for photosynthesis

**Ocean Surface (1983)**
In this masterwork of close observation and technical precision, Vija Celmins uses drypoint engraving, in which the image is incised into a plate using a sharply pointed tool, to recreate wave formations in her limited-edition monochromatic prints.

*the ocean in art*

# lifelike seas

In the 20th century, a new art movement known as Photorealism emerged. This involved artists using a photograph as their primary visual reference to produce images that are so lifelike it is barely believable they have been created by the artist's hand. This art form continues to evolve, with artists using the latest technologies. Among the most astonishing works are the super-realistic seascapes, with their almost tangible waves and sand.

Inspired by Pop Art in the 1960s, the work of the first Photorealists in the 1970s was a commentary on the conflict between art and the explosion of photography in the 20th century. Artists projected the outline of a photograph onto canvas or paper, then demonstrated their extraordinary prowess in mimicking the textures and reflective surfaces of consumer culture icons, such as camping vans and sauce bottles, often using an airbrush. Here were banal subjects lacking any emotional depth presented with the technical skills of an 18th-century portrait artist perfecting the sheen on silks or the texture of fur.

Since the early 1990s, technical advances in digital photography have taken Hyperrealism (as Photorealism is now more commonly known) into different realms. US artist Zaria Forman has flown with NASA survey missions over Arctic and Antarctic seas and has travelled to Greenland to photograph polar ice. In the Maldives, the lowest and flattest region on Earth, she photographed coastlines under threat as sea levels continue to rise. Her message about the beauty and fragility of our world is writ large on vast canvasses of pristine ice masses and tropical waves. Each work captures a photographic moment in time and takes up to 400 hours to recreate.

Latvian-American artist Vija Celmins also works on meticulous renderings of natural phenomena, such as waves, night skies, and the desert floor, often in sombre, dark grey tones. In the late 1960s, she produced highly realistic pencil drawings of the surface ripples in a small area of the Pacific Ocean; more recently, she has replicated those pencil strokes using an engraving tool to create remarkable prints.

❝ My drawings celebrate the beauty of what we all stand to lose. I hope they can serve as records of sublime landscapes in flux. ❞

ZARIA FORMAN, *TED TALKS LIVE*, NOVEMBER 2015

**Maldives No. 11 (2013)**
Zaria Forman conjures up mesmerizing waves lapping on a Maldives shore using only pastel crayons and her fingers and palms. Her giant canvases, produced from her own photographs of vanishing equatorial coastlines and melting ice in polar regions, reflect a planet in peril from climate change.

**Race for the sea**
Two baby loggerhead turtles (*Caretta caretta*) head for the ocean on Florida's east coast. Most marine turtles hatch at night, when land predators are less likely to spot them. They scurry off in the most brightly lit direction – where moonlight or starlight reflects off the water – and away from the shadows of vegetation further inland.

# nesting on beaches

With their paddlelike limbs, marine turtles are superbly adapted to life in water, but their existence is not wholly divorced from the land. Their young, like those of terrestrial tortoises, hatch from hard-shelled, air-breathing eggs. The female turtles bury them on sandy beaches, but the strategy is not without risk. On exposed shores, hatchlings must brave land predators in the brief – but perilous – scamper to the sea. Females offset the certain loss by producing huge clutches, typically containing more than 100 eggs.

## TEMPERATURE AND SEX

The sex of a turtle embryo is determined by temperature: above 29°C (84°F), sex organs are more likely to become ovaries, producing females; below this, they develop as male testes. Buried eggs incubate as sunshine warms the sand, but the temperature inside the nest is not uniform – eggs buried deeper are slightly cooler. This temperature variation ensures that roughly equal numbers of females and males hatch.

Warmer top of nest produces more females

Cooler, deeper nests produce more males

Males more likely to hatch from eggs at bottom

**SHALLOWER NEST**

**DEEPER NEST**

Each turtle crawls beyond the reach of the tide to lay its eggs; the entire beach visit lasts about an hour

### Natal rookery

Female olive ridley turtles (*Lepidochelys olivacea*) return in their thousands to the beach nesting grounds of their birth, called rookeries. They use their hind legs to dig a hole in which to lay eggs, then bury the clutch and return to the sea.

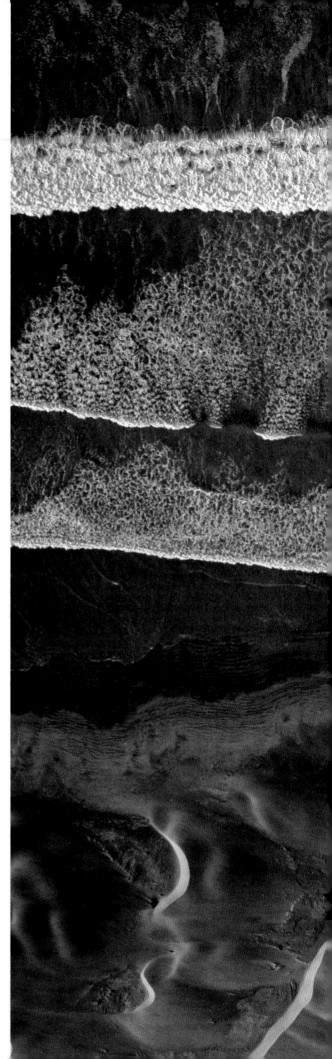

sandy beaches

**Sculpted by the wind**
Barchan dunes form when the wind,
coming predominantly from one direction,
blows the sand into distinctive crescent-
shapes as on Isla Magdalena, on the
Pacific coast of Baja California, Mexico.

# beaches and dunes

Beaches are zones where sediment in the form of sand –
created by the action of rivers, cliff erosion, and transport from
deep water towards the shore – accumulates before eventually
being carried back to the sea by wave action. Beaches lose sand
to the sea by the action of strong offshore rip-currents and
winter storms. Sand may also be transported inland by wind
to form dunes. The most important force for sand transport is
persistent longshore drift, which drives a conveyor belt of supply
and removal. The colour of sand varies according to the type of
rock: grey-black from volcanic rocks; glistening white from coral
and limestone; or golden, mainly comprising quartz coated with
iron oxide.

## TYPES OF BEACH

Pocket and embayed beaches nestle in the more protected regions between
headlands. Broad sandy beaches with gentle gradients are often flanked by dunes.
Where wave energy is greater, sand is washed away and steep pebble beaches are
built up. Where waves approach a shoreline obliquely, they move sand by a process
of longshore drift and create sand spits, barrier islands, and sheltered lagoons.

Lagoon

Longshore drift

Soft rock

Hard rock

Embayed beach

Pocket beach

Sand spit

Wave direction

> Aloft all hands, strike the top-masts and belay;
> Yon angry setting sun and fierce-edged clouds
> Declare the Typhoon's coming.

J.M.W. TURNER, UNTITLED POEM (1812) ACCOMPANYING *THE SLAVE SHIP*, 1840

## The Sea of Ice (1823–24)

The seascapes of German artist Caspar David Friedrich evoke awe in the natural world and a sense of human irrelevance in the face of an almighty power. Here a ship is crushed in its ice tomb, while the jagged ice is driven into pinnacles against the infinite blue of the sky.

*the ocean in art*

# drama on the seas

In the Romantic movement, which spanned most of the 19th century, inspiration, originality, and imagination were key. Artists dispensed with neoclassical rules and norms to focus on their emotional response to the natural world and to the plight of individuals. Many iconic seascapes of the period, featuring towering waves and apocalyptic skies, tell contemporary stories about human endeavour, disaster, and oppression.

The 19th-century French poet and art critic Charles Baudelaire described Romantic art as "precisely situated neither in choice of subject, nor in exact truth, but in a way of feeling." For some artists, the freedom to work outdoors (made easier by the invention of paint tubes) fuelled a deeper engagement with land and sea. In studios, experimentation with brushwork, colour, and form fostered a range of individual techniques.

Russian marine artist Ivan Aivazovsky conjured a sense of the ocean's sublime magnitude with vast canvasses saturated with colour and light. In *The Ninth Wave* (1850) the rising Sun lights a huge wave that is threatening to engulf the survivors clinging to the wreckage of their ship.

In a period when European colonialists were exploiting overseas territories, and scientists and explorers were sailing to remote regions, outrage and adventure surfaced in Romantic art. French painter Théodore Géricault's *The Raft of the Medusa* (1819) is a striking recreation of the true story of crew and passengers abandoned by officers when the French frigate *Méduse* ran aground on its way to Senegal. In this work, the deceased slip from a poorly made raft, survivors cling to life, and hope of rescue is encapsulated in a makeshift flag waved towards a brightening sky. More fancifully, English artist Sir Edwin Landseer imagined the outcome for Sir John Franklin's failed attempt to find a Northwest Passage through Arctic seas. In his *Man Proposes, God Disposes* (1864) polar bears invade the ice-trapped wreckage to feed on human bones.

A large proportion of the 19,000 watercolours and oils painted by the English artist J.M.W. Turner were seascapes. Over a lifetime he perfected washes of paint, textures, and bursts of light to create the natural phenomena of rain, fog, storm, sunrise, and sunset. He based his painting and poem *The Slave Ship* on the true story of the captain of the *Zong*, who ordered his crew to throw sick and dying slaves into the sea to guarantee insurance payments – a common practice in slave shipment.

## The Slave Ship (1840)

In J.M.W. Turner's celebrated work, the churning waves look as though they are on fire and the ship ploughs ominously into the storm as the captain carries out his evil mission. Turner's full title, *Slavers Overthrowing the Dead and Dying – Typho[on] Coming On*, supplies the narrative for the carnage in the foreground: the flailing, shackled limbs of submerged slaves who have been thrown into the shark-infested sea.

# american crocodile

The American crocodile (*Crocodylus acutus*) exists in a range of habitats from freshwater rivers to brackish swamps and coastal waters, and is found in regions as far apart as Florida, USA and the Caribbean, as well as off both the Atlantic and Pacific coasts of South America.

The American crocodile is one of the largest of the crocodile species; adults have an average length of 4.3 m (14 ft). Like all crocodiles, it has salt-secreting glands on its tongue that removes excess salt, helping it survive in higher salinities.

The American crocodile has an especially large range of tolerance: most populations live in brackish lagoons, mangroves or estuaries, but others live in freshwater rivers and reservoirs much further inland. The largest Caribbean population – in the Dominican Republic – even succeeds in a salt lake with water that can be three times saltier than the ocean. A wide distribution across the American tropics is testament to the apparent adaptability of the American crocodile. The fact that it has colonized islands, including coral atolls, is proof that it often swims far out to sea, possibly aided by tidal currents.

While an American crocodile's jaws can exert hundreds of kilograms of pressure when it bites, easily crushing an adult turtle shell, they can also be very gentle. Females can pick up and carefully squeeze unbroken eggs to help the young crocodiles hatch, and they often carry their young in their mouths to transport them to the water. A crocodile's cone-shaped teeth are designed for grabbing and holding prey but not for chewing. They are constantly being replaced, and an animal may go through 8,000 teeth during its lifetime.

### Adapted for hunting

A crocodile senses prey by smell and via sensory bumps on the sides of its jaws that detect vibrations. They can swim open-mouthed and have excellent vision; a transparent third eyelid protects the eye when the animal is submerged.

### SURVIVING UNDERWATER

A crocodile's mouth is not watertight, so to prevent water flooding its lungs and gut when underwater, a palatal valve – a large flap behind the tongue – seals the entrance to the windpipe and gullet. At the same time, smaller nostril valves close the nose. In this way, a crocodile can stay submerged without breathing for over an hour, and it can even open its mouth wide enough to grab or manipulate prey. The raised position of the open nostrils also allows the crocodile to breathe air while swimming or hanging just below the water surface.

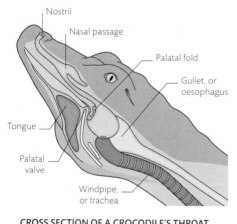

CROSS SECTION OF A CROCODILE'S THROAT

Nostril

Nasal passage

Palatal fold

Gullet, or oesophagus

Tongue

Palatal valve

Windpipe, or trachea

# beachcombing

A sandy beach can provide a good living, even for animals that never enter the water. Each tide brings a deposit of natural flotsam, in the form of seaweed. This, in turn, attracts swarms of flies, beetles, and tiny crustaceans: food for shorebirds, such as plovers. Up beyond the ocean's reach, among the sparsely vegetated dunes, birds build their nests with sticks, stones, shells, and anything else they might collect as they forage along the water's edge.

**Life above the tides**
Along the southern Australian coastlines, the hooded plover (*Thinornis cucullatus*) prefers beaches strewn with plenty of seaweed. Its legs are too short to wade for its food, so it sticks to the strandline when foraging for invertebrates.

**Large eyes** help the bird to spot small prey among the tideline debris

**Disguised egg clutch**
The bird's speckled eggs resemble the surrounding pebbles. The nest may be lined with seaweed and shell fragments for additional camouflage. Once hatched, the sandy-coloured chicks blend in too.

**The hooded plover's nest** is a shallow depression called a scrape

## NESTING TOGETHER

Even when partly concealed by the spoils of beachcombing, the nests of hooded plovers are vulnerable to predation. Being territorial breeders, hooded plovers prefer to nest over 20 m (66 ft) apart. Ground-nesting birds often breed in loose colonies, typically with mixed species for extra vigilance, and hooded plovers may be joined by pied oystercatchers (*Haematopus longirostris*).

**PIED OYSTERCATCHERS**

**Seaweed fronds** detached by wave action attract kelp flies and other decomposers, providing a ready source of food for birds

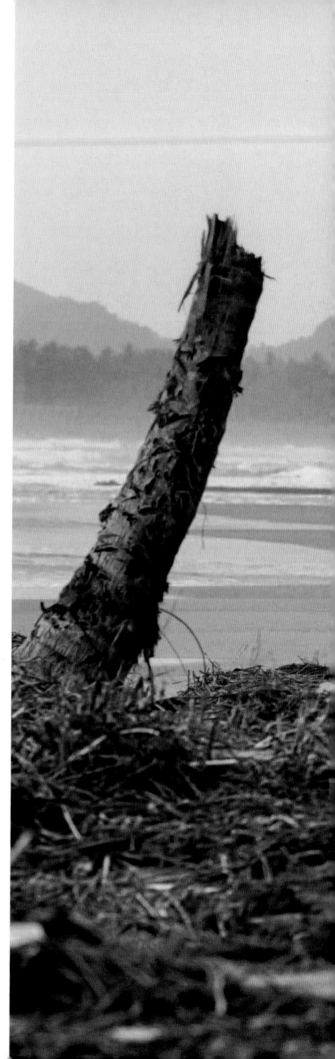

**Coastal devastation**
The aftermath of the 2004 tsunami that ravaged the west coast of Aceh, Indonesia, pictured in 2005, testifies to the destructive power of this ocean event over wide tracts of the formerly tree-covered shoreline.

# tsunamis

Tsunamis are unusual waves of extreme wavelength (up to 200 km/125 miles) that can travel across an entire ocean at a speed of up to 800 kph (500 mph). A common cause of tsunamis is the sudden vertical movement of the seafloor due to earthquakes or giant submarine slides under the sea. They may also be caused by events on land that deposit vast amounts of material into the sea, such as avalanches, landslides, volcanic eruptions, or the calving of large icebergs from ice-sheets or glaciers. In each case, the enormous volume of ocean water displaced gives a tsunami its huge power and velocity. Tsunamis can originate in any ocean but are most common in areas of frequent volcanic activity.

## FROM UNDERSEA SHOCK TO COASTAL DESTRUCTION

A tsunami often starts with a sudden displacement of the ocean floor, which causes a large movement of water that produces a long, low, almost imperceptible wave in the open ocean. The wave's height increases dramatically as it crosses the shallow continental shelf and sweeps towards land, forming a giant wall of water – up to 30 m (100 ft) high – that surges inland, wreaking havoc along the coast.

Water elevated above fault

Direction of wave

Wave reaches maximum height near coastline

Shockwave created by sudden displacement of seafloor along fault line

Wave compressed as it reaches shallower water

# estuaries
## and mudflats

**The world's muddy shorelines** and estuaries are home to animals that can burrow in sticky mud and to plants and animals that can withstand the constant changes in salinity that accompany the rise and fall of the tide.

# living in sediment

Among the invertebrate animals that live in mud are many kinds of bivalve: molluscs with a two-part shell, such as clams and cockles. They use their gills to filter particles of food, including the organic detritus and microorganisms that typically enrich the seafloor. Many bivalves use their muscular foot to burrow into the substrate, away from surface predators. Surrounded by sediment, these molluscs then draw water through their body to extract both oxygen and nourishment.

The foot can extend up to 6 cm (2.5 in) as it burrows

**Burrowing bivalve**
A prickly cockle (*Acanthocardia echinata*) pushes its muscular foot into the sand. The foot then expands near its tip to achieve traction, before pulling the rest of the cockle's body beneath the surface.

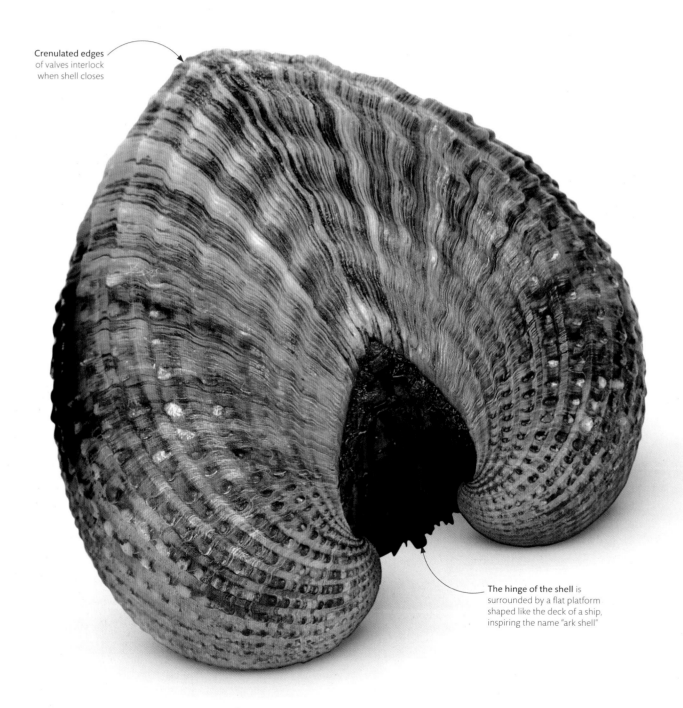

Crenulated edges of valves interlock when shell closes

The hinge of the shell is surrounded by a flat platform shaped like the deck of a ship, inspiring the name "ark shell"

## SIPHONING WATER

Water does not easily circulate through sticky mud, so many burrowing bivalves use siphons to draw water down from the surface. The siphons extend when the valves open. Those with the longest siphons can burrow the deepest. Cockles, with short siphons, and ark shells, which lack siphons, live closer to the surface.

**CROSS SECTION OF COCKLE FILTER-FEEDING IN MUD**

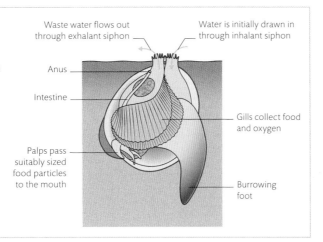

Waste water flows out through exhalant siphon

Water is initially drawn in through inhalant siphon

Anus

Intestine

Gills collect food and oxygen

Palps pass suitably sized food particles to the mouth

Burrowing foot

## Hinged shell

Like other bivalves, the Indo-Pacific blood clam (*Tergillarca granosa*) from tropical coastal muds has a shell made of two parts, or valves, connected by a hinge of tough ligament. The valves are pulled shut by muscles when danger threatens. As these muscles relax, the shell gapes open, enabling the mollusc to extend its muscular burrowing foot and circulate water for filter-feeding.

**Fleshy foot** is powered by multiple muscles that extend the foot or move it from side to side when burrowing

**Ripples in the sand**
As the tide retreats on the Scottish Hebridean island of Eigg, the action of the water moulds the fine sand into distinctive ripples. The metallic sheen of the sand is the result of its high content of volcanic material.

# tides

The regular – usually twice daily – rise and fall of sea level known as the tide has been a constant phenomenon since the oceans first formed four billion years ago. Tides are waves with long wavelengths that sweep around bodies of water (oceans and seas), moving up and down low-lying coastal zones, with dramatic effects on life in the shallows. High tides are the crests of the waves, and low tides are the troughs. The difference in height between high and low tide is termed the tidal range. This varies from less than 1 m (3 ft) to just over 16 m (53 ft). Tides are caused by the combined gravitational forces of the Earth–Moon and Earth–Sun systems, and by the spin of Earth in these force fields.

## MONTHLY CYCLES

When the Sun and Moon are aligned (at full Moon and new Moon), the two force fields act in the same direction, reinforcing each other – the spring tide. When they are at right angles (at half-Moon), the combined tide-generating forces are at a minimum – the neap tide. The horizontal component of gravitational force on a body of water is the most significant. Each ocean or sea has its own slightly different tidal pattern.

# seabed stinger

While some sea animals use venom to attack prey, many fish sting only in self-defence. Stingrays, which are cartilaginous fish related to sharks, spend their time resting on the seafloor and hunting prey in the sediment. In this position, with their mouth on the underside, they are vulnerable to attack from above. If they perceive a direct and inescapable threat, for example when trodden on, they lash their barbed tail upwards and deliver their venom.

Sand is whipped up by stingray feeding

**Feeding in the sand**
When a stingray hunts, it fans its fins and squirts jets of water from its mouth in order to dislodge its prey (small fish and invertebrates) buried in sediment on the seabed.

Pointed fin-tips give a diamond-shaped outline

## WHIP AND BARB
The narrow tail of a stingray can be moved at whiplike speed. When the stingray strikes, the tail flicks forwards over the head to stab the target with a serrated spine, which produces venom and is encased in a sheath. The spine may break, but it can regrow within several months.

Serrated edge of spine

Sheath

Spiracle

Pair of venom glands in two grooves in spine underside

Whiplike tail

Eye

Winglike pectoral fin

**STINGRAY (UPPER SIDE)**

Light coloration of underside provides camouflage, making the stingray harder to see from below

## Lying low

Like all rays, the southern stingray (*Hypanus americanus*) is dorsoventrally flattened, with a low profile suited to life on the seabed. The mouth is on the underside to enable it to feed in the sediment. To prevent sand clogging the gills, water is taken in via spiracles on top of the head. These draw in water, which is passed over the gills and expelled from its gills on the underside.

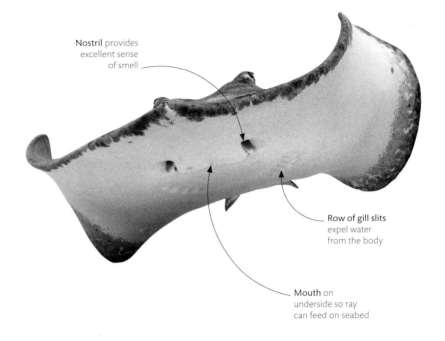

**Nostril** provides excellent sense of smell

**Row of gill slits** expel water from the body

**Mouth** on underside so ray can feed on seabed

**Small spines** on midline

**Spiracles** on top of head enable stingray to breathe when lying on seabed

**LETHAL VENOM**
Stonefish are the world's most venomous fish, although they use their sting defensively rather than for hunting. Stored in a pair of venom glands, the poison is delivered via 13 modified dorsal fin spines. These act like hypodermic needles to inject a cocktail of toxins that attack a victim's muscles, nerves, blood cells, and cardiovascular system, causing agonizing pain, shock, paralysis, and severe tissue damage.

Opening obstructed by sheath

Pressure applied to spine from above triggers sting

Sheath

Venom duct in spine

Pair of venom glands

Connective tissue

**PRIMED**

Venom released

Punctured sheath

Venom in glands activated

**DELIVERED**

**Engulfed in an instant**
The chief hunting aids of the reef stonefish (*Synanceia verrucosa*) are its prominent eyes and large mouth, the outlines of which are disguised by highly textured skin. When the stonefish spots its prey, such as smaller fish or crustaceans, its mouth opens in less than a fiftieth of a second, creating suction that prey are powerless to resist.

# concealed danger

Scorpionfish and stonefish are ambush predators, but unlike stargazers (see pp.232–33) and anglers, which use a lure to tempt prey closer, they rely purely on camouflage and an ability to remain perfectly still for hours at a time. Some scorpionfish and stonefish species are able to change colour to match their background, and each fish appears to select resting places that best match their individual coloration.

**Skin** is highly textured and sometimes carries algae, which may attract herbivorous prey closer to the stonefish

**Dorsal fin** contains spines that deliver venom

**Hiding in plain sight**
Lurid pink, orange, and purple are not obvious camouflage colours, but set against the complex background of a coral reef, a reef stonefish can remain invisible to prey and predators until it moves. Despite their camouflage and venom, small stonefish are occasionally eaten by predators such as sea snakes, sharks, and rays.

**Large pectoral fins** and tapered tail are often shuffled under sediment or tucked into crevices, making stonefish even harder to spot

estuaries and mudflats | 86 • 87

**Prominent eyes** increase field of vision

**Large mouth,** which is fringed with warts and tassels, has an upward-protruding jaw; the mouth expands rapidly to engulf prey above

**Elongated snout,** which resembles the blade of a saw, is an extension of the skull

**Skeleton** is made of cartilage instead of bone

**Saw-teeth** are evenly sized, flat, and peglike, with a sharp point

**Nostrils** positioned on underside of head

**A side-to-side** sweeping motion is used to detect buried prey

**Sweep and slash**
A sawfish uses its saw-like snout in several ways: to scan the seabed for electrical signals from prey hidden in the sand, to swipe at and disable its prey in open water, and to thrash at an assailant or sometimes at other sawfish when competing for food.

# sensing electricity

Sharks, rays, and skates are known for their extraordinary sense of smell, but they also possess an electroreceptive sense located in pinhole-sized pits on the snout. These pits are especially abundant on the enlarged snouts of hammerhead sharks, sawsharks, and sawfish. The latter are actually rays and only distant cousins of sawsharks. Their similar-looking "saws" are examples of convergent evolution, whereby different species have evolved similar traits in order to adapt to the same environment.

**Sawfish teeth**

Like all sawfish, the smalltooth sawfish (*Pristis pectinata*, shown here as an X-ray) has a long, narrow snout (rostrum) and two sets of teeth. The most obvious are the large, robust, sharply pointed saw-teeth, which are found all along the snout. In addition, the mouth contains a set of tiny, blunt teeth, arranged in 10–12 rows, which are used for crushing prey.

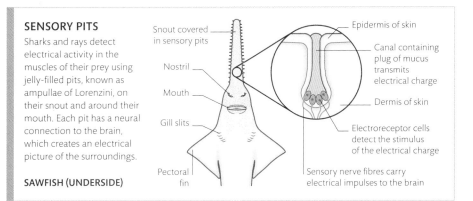

**SENSORY PITS**

Sharks and rays detect electrical activity in the muscles of their prey using jelly-filled pits, known as ampullae of Lorenzini, on their snout and around their mouth. Each pit has a neural connection to the brain, which creates an electrical picture of the surroundings.

**SAWFISH (UNDERSIDE)**

Snout covered in sensory pits

Nostril

Mouth

Gill slits

Pectoral fin

Epidermis of skin

Canal containing plug of mucus transmits electrical charge

Dermis of skin

Electroreceptor cells detect the stimulus of the electrical charge

Sensory nerve fibres carry electrical impulses to the brain

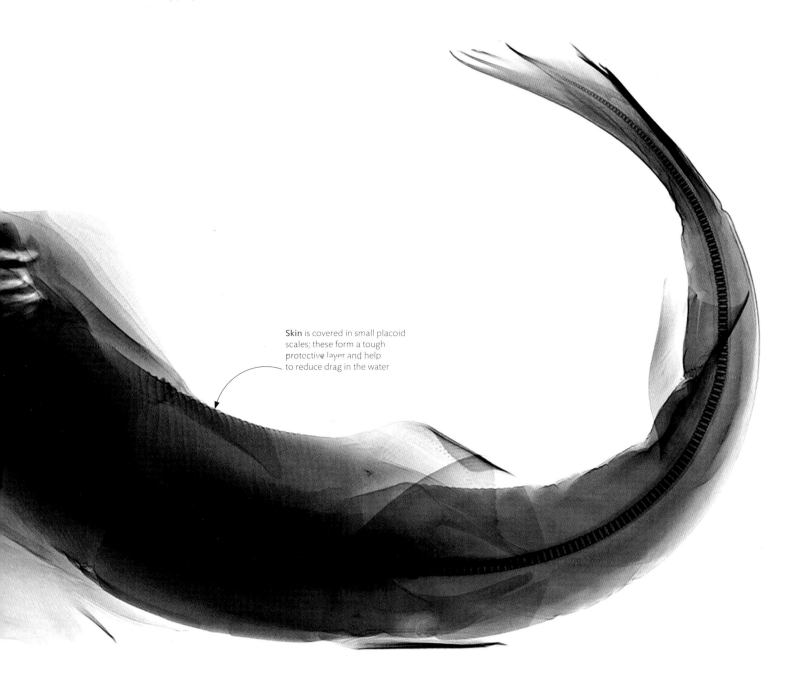

**Skin** is covered in small placoid scales; these form a tough protective layer and help to reduce drag in the water

# migrating to breed

Many fish return to the place where they began life in order to breed. In doing so, they increase the likelihood that their offspring will have similar success. For salmon, which spend most of their adult lives at sea, the return journey to their home rivers and streams is particularly challenging. First, they undergo physiological changes to enable them to reacclimatize to fresh water. During the upstream journey, which can cover hundreds or even thousands of kilometres, they have to battle swift currents and leap waterfalls, usually without feeding. The strain of the journey combined with hormonal changes lead to programmed death within days of spawning.

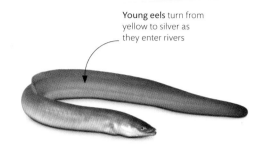

**Young eels** turn from yellow to silver as they enter rivers

**Migrations in reverse**
While salmon are anadromous, migrating upstream to spawn and die, freshwater eels are the opposite (catadromous): they begin life at sea, migrate to fresh water, and return to sea to breed.

## BODY CHANGES

Salmon undergo many changes in their life history. Newly hatched "fry" develop camouflage markings in the form of vertical bars (the "parr" stage), and as the fish prepare for life at sea (the "smolt" stage) they turn silver. The oceanic phase lasts up to five years, then salmon change colour again – to brown, red, or green, depending on the species – before they return to fresh water. Breeding males of several species undergo changes in shape, most notably the pink salmon (right).

**OCEANIC PHASE**

Silvery scales and a torpedo-shaped body (both sexes)

**BREEDING FEMALE**

Turns brown; body shape remains similar

**BREEDING MALE**

Develops distinct hump and hooked jaw (kype)

### Homing instinct

Salmon rely on two main senses to navigate their way back to their spawning grounds. An ability to detect Earth's magnetic field leads them to their home river, then they use their acute sense of smell to find the gravel bed where they hatched. Here, pink salmon (*Oncorhynchus gorbuscha*) in Alaska, US, are heading upriver to breed.

# mud probing

Many shorebirds, including members of the sandpiper family, breed around the Arctic tundra. As the days shorten and insect food dwindles, they fly south to overwinter on intertidal mudflats around the world. Here they enjoy a rich diet of mud-dwelling invertebrates, which helps to fuel their long-distance journeys and sustains them through the winter months. Each species is equipped with the perfect tool – a probing bill – to secure a meal.

## BILLS FOR PURPOSE

Having different bills with different uses reduces competition for food among shorebirds – even between related species, such as sandpipers. Long, pointed bills – curved or straight – can penetrate deep into thick sediment in search of invertebrate food. Shorter bills are more useful for foraging near or at the surface of the intertidal mud.

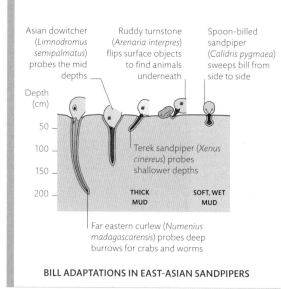

Asian dowitcher (*Limnodromus semipalmatus*) probes the mid depths

Ruddy turnstone (*Arenaria interpres*) flips surface objects to find animals underneath

Spoon-billed sandpiper (*Calidris pygmaea*) sweeps bill from side to side

Depth (cm)

50
100
150
200

Terek sandpiper (*Xenus cinereus*) probes shallower depths

THICK MUD

SOFT, WET MUD

Far eastern curlew (*Numenius madagascarensis*) probes deep burrows for crabs and worms

**BILL ADAPTATIONS IN EAST-ASIAN SANDPIPERS**

## Switching diets

After catching fly larvae and beetles during the Canadian summer, the short-billed dowitcher (*Limnodromus griseus*) migrates southwards and repurposes its bill for probing, as it hunts for mud-burrowing worms, molluscs, and crustaceans along coastlines of the American tropics.

Shorter bill may be better for catching insects

**Age-related bills**
Like others of the sandpiper family, the insect-eating chicks of the black-tailed godwit (*Limosa limosa*) hatch with proportionately shorter bills than those of mud-probing adults.

The horny sheath around the bill is honeycombed with cavities containing pressure sensors, which enable the bird to detect the movements of buried prey

**Slender head** may be plunged part-way into the mud when foraging, helping the bill to reach deeper for burrowing invertebrates

**Brown-grey plumage** may offer some camouflage from predators when viewed against the ground, both when breeding in subarctic meadows and when feeding on mudflats

**Flying the flag (1660–80)**
Celebrating trade in the Dutch Golden Age, this unnamed work by marine artist Abraham Storck depicts merchant ships and galleons at anchor in a tranquil fantasy port. It is a counterpart to another Storck work, *Dutch Ships for a Port in the Mediterranean Sea*.

**A Dutch Ferry Before a Breeze (1640s)**
A red-sailed ferry carrying passengers and goods across a blustery bay showcases Simon de Vlieger's draughtsmanship and command of the atmospheric effects of waves and weather.

*the ocean in art*

# the dutch golden age

Riding the crest of a wave in the 17th century, the Dutch rose to great power through their mastery of the sea. Their unparalleled boat- and shipbuilding skills for transportation, trade, and war were matched by a prolific output of great art in every genre. The works of marine artists were especially popular, capturing the might of galleons, the passage of small vessels on inland waters, and the atmospheric effects of weather, sea, and sky.

In 1640, merchant and traveller Peter Mundy counted 26 trade ships arriving at the Dutch Texel Inlet from distant shores and concluded that nowhere in the world came near to it for traffic and commerce by sea. Art, he noted, was a prerequisite in even the poorest homes and in the stalls of butchers, bakers, and blacksmiths. Current-day estimates suggest that between 1640 and 1660 Dutch artists produced more than 1.3 million pictures. Paintings of huge fleets and merchant ships in foreign ports reflected Dutch maritime and economic dominance. Major works, such as Abraham Storck's *The Four Days' Battle* (1666), showing fighting in the Anglo-Dutch war, reflected constant naval activity to keep trade routes open. Other artists, including Jan van Goyen, Ludolf Backhuysen, and Willem van de Velde the younger, depicted coastal sailing and the quiet harbours, fishing inlets, and winter skating intrinsic to Dutch domestic life.

66 As For the art off Painting and the affection off the people (of the Netherlands) to Pictures, I thincke none other goe beeyond them... 99

PETER MUNDY, *THE TRAVELS OF PETER MUNDY IN EUROPE AND ASIA*, 1608–1667

**In the pink**
As with flamingos and the scarlet ibis, the pink coloration of the roseate spoonbill (*Platalea ajaja*) comes from carotenoid pigments that are found in its food. The intensity of colour varies according to age, season, and location, and the availability of carotenoid-rich prey.

**Head and neck** become bald in adult birds

**Throat** and oesophagus are elastic to accommodate prey, which is swallowed whole

**Plumage** ranges from white to deep crimson

# sweeping for food

For birds feeding in murky water, it is often difficult to see beneath the surface. As a result, several species have evolved a sensitive bill that enables them to detect food in shallow waters, without relying on their sense of sight – an adaptation that also allows them to continue feeding at night. All six species of spoonbill hunt by "tactolocation", detecting small fish, shrimps, and aquatic insect larvae and plant matter in lagoons, pools, and estuarine waters up to 30 cm (12 in) deep.

**Orange bill** of young spoonbills fades to grey in adult birds

**Early days**
Young spoonbills start life with a short, straight beak. This elongates rapidly in the days after hatching, begins to broaden at the tip after one to two weeks, and attains the characteristic adult shape after about one month.

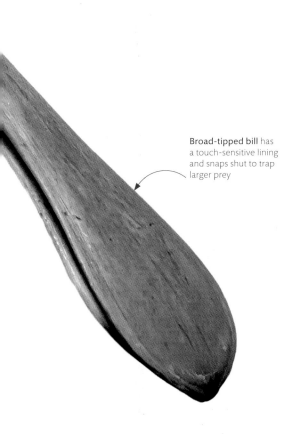

**Broad-tipped bill** has a touch-sensitive lining and snaps shut to trap larger prey

## FEEDING TECHNIQUE

A spoonbill feeds in shallow water, sweeping its partially open bill in a side-to-side or circular motion to create a vortex. After dislodging food from the bottom, it traps the catch between its broad mandibles, lifts its head, and claps its bill shut, crushing its prey before swallowing.

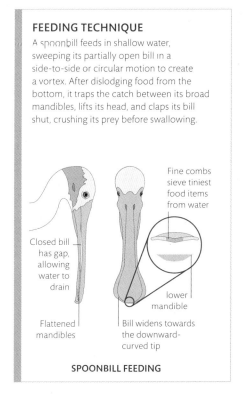

Closed bill has gap, allowing water to drain

Flattened mandibles

Fine combs sieve tiniest food items from water

lower mandible

Bill widens towards the downward-curved tip

**SPOONBILL FEEDING**

**Tail** helps the osprey to manoeuvre into position, so that the bird can accurately hit the target

**Wings** pull back in W-shape, helping the osprey to cut through the air when diving towards the water's surface

**Feet** swing forwards, ready to grab the fish, just before the bird impacts the water

**Preparing for splashdown**
Ospreys plunge from heights of 40 m (130 ft) on long, narrow wings. This wing shape is ideal not only for diving but also for hovering over large bodies of water while selecting a suitable target.

# fishing with feet

Raptors – birds of prey, such as eagles and hawks – typically catch their quarry with talons that pierce the prey's flesh. The osprey (*Pandion haliaetus*) is the only daytime raptor that feeds almost exclusively on fish. It hunts like other raptors, meaning that it dives feet first – but it does so mainly in saltwater habitats. Once its target is properly gripped, the osprey is capable of flying back to its perch carrying a fish half its body weight.

## CATCHING SLIPPERY PREY

Raptors usually have an arrangement of toes with three pointing forwards and one pointing back – like most types of bird. This helps with perching as well as gripping prey. But slippery, wriggling fish need a firmer grip, so the osprey has a reversible outer toe, allowing two talons to hook either side of a catch. Smaller fish can be held with one foot, but heavier ones are secured using both feet, one in front of the other.

Normal position of outer toe

Outer toe twists backwards to get a firmer grip

**OSPREY FEET**

**Bill** has a long, sharp hook that is able to penetrate the tough, scaly skin of large fish

**Dense, oily plumage** keeps the feathers waterproofed when diving for fish

**Toes** carry spiny pads on their undersides, which enable the feet to cling tightly to a fish

### Coastal raptor
Despite having long legs to help reach for fish, an osprey's natural buoyancy means it cannot penetrate more than about 1 m (3 ft) under water. It therefore catches surface-dwelling fish in shallow bays and along shorelines. This technique is successful worldwide: the osprey is one of the most wide-ranging of all bird species, found in the Americas, Africa, Eurasia, and Australasia.

# mangroves
## and salt marshes

**These tidal habitats** are dominated by trees and other true plants that have adapted to life in salt water. They also protect shorelines from coastal erosion and provide homes to diverse communities of other living things.

# growing in salt marshes

Plants that grow on land or rooted in fresh water rely on salts in their tissues to pull water up into their leaves: water tends to seep towards higher salt concentrations by a process called osmosis. Plants of intertidal marshes, such as sea-blites, are specially adapted to the saltier conditions, typically with fleshy leaves that accumulate the extra salt needed to keep drawing in water.

Maturing shoots turn violet-red with betacyanin, the same pigment found in beetroot

**Bicoloured shoots**
Sea-blites, such the Australian *Suaeda australis*, turn from green to red as they build up betacyanin – a pigment that protects cells in the salty conditions.

## GAINING AND LOSING WATER

Plants adapted to salty conditions are called halophytes. By concentrating salt in their tissues – at higher levels than in their surroundings – they absorb water by osmosis, from lower to higher salt concentration. Conventional land plants, known as mesophytes, have lower salt concentrations than sea water, so they lose water by osmosis in salty conditions.

**SALT CONCENTRATION**

Normal soil   Inside mesophyte
Salty soil    Inside halophyte

Plant gains water by osmosis

Plant gains water by osmosis

Plant loses water by osmosis and wilts

Plant gains water by osmosis

**MESOPHYTE ON NORMAL SOIL**

**MESOPHYTE ON SALTY SOIL**

**HALOPHYTE ON SALTY SOIL**

**Salt marsh annual**
In China's Liaohe River delta, conditions are too wet and salty for most types of plant to survive. But the Eurasian sea-blite (*Suaeda salsa*) thrives in abundance. It grows from seed each year, and during autumn its shoots gradually transform the shoreline into a spectacular red "beach".

**Colourful colonists**
Colonizers of mangrove roots include bluebell tunicates (*Clavelina*). Where sunlight can reach, photosynthesizing green algae thrive alongside them.

The tunicates' bottle-shaped bodies draw in water to filter out food particles

A colony of tunicates arises because each individual can produce neighbouring clones

# colonizing roots

Along muddy shorelines around the tropics, mangrove roots growing like pillars in the water often provide the only solid surface in the intertidal habitat. As a result, they are quickly colonized by opportunistic algae and animals. Competition for space is intense. For the mangrove, these underwater "gardens" are a mixed blessing: they nourish the roots with their waste but also smother the pores through which the roots breathe.

**Underwater garden**
Submerged under nutrient-rich waters at high tide, these mangrove roots are all but hidden by plankton-feeders such as tentacled polyps and bottle-shaped tunicates. Colonists that grow faster may have the advantage, but some can produce chemicals that repel their neighbours.

**TUNICATE LIFE CYCLE**
Tunicates, like many ocean invertebrates, have a life cycle that alternates between being attached and being free-swimming. As sedentary filter-feeders, they trap food particles from sea water circulating through their soft bodies. Their non-feeding larvae are supported by a firm rod called a notochord. This structure is anatomically equivalent to a backbone – evidence that tunicates are more closely related to vertebrates than to other invertebrates.

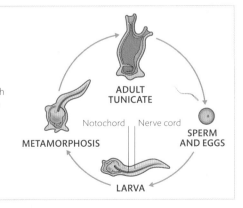

ADULT TUNICATE

Notochord   Nerve cord

SPERM AND EGGS

METAMORPHOSIS

LARVA

# growing on stilts

A plant that grows into a tall tree benefits from reaching more light than its shorter neighbours. But the extra weight involved in scaling such heights needs strong support. On muddy tropical shores, salt-tolerant trees called mangroves manage on the soft ground by distributing their weight over a wide area with stiltlike roots or with buttress roots that spread horizontally along the ground. This stops the trees from toppling, even when their bases are submerged by the incoming tide. The high, arching stilts also collect vital oxygen, enabling the trees to thrive where the sludgy ground is not easily aerated and keeping the root tissues alive far below the surface.

**Salt crystals** form when saline water secreted by leaf glands evaporates

**Coping with salt**
Mangroves can tolerate high levels of salt in their tissues. Some species exude excess salt from their leaves, which become coated with white salt crystals.

## ROOTS THAT BREATHE

In order to distribute oxygen, mangrove roots are honeycombed with air-filled spaces. Air enters the roots at high tide through pores called lenticels, but the roots' shape varies between species. Unlike the stilt-rooted *Rhizophora* mangroves, the roots of *Bruguiera* emerge from the mud as a series of loops, whereas *Avicennia* mangroves reach the air through pneumatophores – root tips that grow upwards from the mud's surface, effectively acting as snorkels.

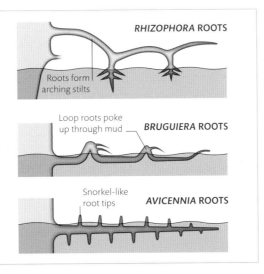

RHIZOPHORA ROOTS

Roots form arching stilts

Loop roots poke up through mud

BRUGUIERA ROOTS

Snorkel-like root tips

AVICENNIA ROOTS

### Shoreline thicket

The stilt-based architecture of mud-growing mangrove trees provides a unique forest habitat along tropical coastlines around the world. In southern Asia, the arching roots of the red mangrove (*Rhizophora mangle*) are home to small foraging animals as diverse as crabs, mudskippers, and lizards – and even to monkeys and birds.

Brown-coloured symbiotic algae are concentrated in the oral arms; by swimming upside-down, the jellyfish exposes the algae to sunlight

# living upside-down

Living in mangrove-fringed lagoons and in seagrass beds – often gathering by the hundreds – is an unusual jellyfish that lives upside-down, with its stinging arms waving upwards in the water. Settled on the bottom in the sunlit shallows, it relies on brown-coloured algae peppering its flesh to make food by photosynthesis. Food made in this way supplements the nutrients the jellyfish derives from its captured prey.

**Seabed jelly**
The Indo-Pacific upside-down jellyfish (*Cassiopea andromeda*) pulsates its bell and flickers its arms in the clear, warm water. The arms are multipurpose: as they move back and forth through the water, they absorb oxygen, trap food, and expose their food-making algae to the sunlight.

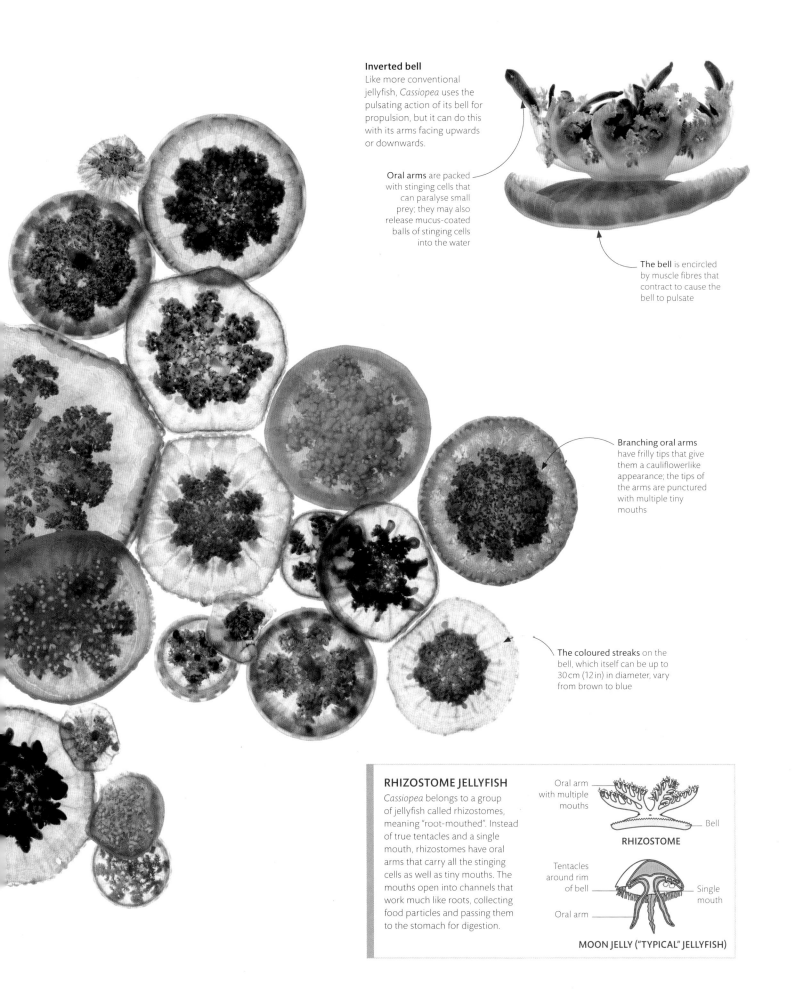

**Inverted bell**
Like more conventional jellyfish, *Cassiopea* uses the pulsating action of its bell for propulsion, but it can do this with its arms facing upwards or downwards.

**Oral arms** are packed with stinging cells that can paralyse small prey; they may also release mucus-coated balls of stinging cells into the water

**The bell** is encircled by muscle fibres that contract to cause the bell to pulsate

**Branching oral arms** have frilly tips that give them a cauliflowerlike appearance; the tips of the arms are punctured with multiple tiny mouths

**The coloured streaks** on the bell, which itself can be up to 30 cm (12 in) in diameter, vary from brown to blue

### RHIZOSTOME JELLYFISH

*Cassiopea* belongs to a group of jellyfish called rhizostomes, meaning "root-mouthed". Instead of true tentacles and a single mouth, rhizostomes have oral arms that carry all the stinging cells as well as tiny mouths. The mouths open into channels that work much like roots, collecting food particles and passing them to the stomach for digestion.

Oral arm with multiple mouths

Bell

**RHIZOSTOME**

Tentacles around rim of bell

Single mouth

Oral arm

**MOON JELLY ("TYPICAL" JELLYFISH)**

# living fossil

Horseshoe crabs appeared in the oceans half a billion years ago, when the jointed-legged arthropods were diversifying via evolution into insects, arachnids, and crustaceans. Ancient, fossilized horseshoe crabs are very similar in shape to those alive today, suggesting that these animals have changed little over millions of years of evolution. Despite their name and their hard shell, they are more closely related to spiders than to true crabs.

**Ancient survivor**

The mangrove horseshoe crab (*Carcinoscorpius rotundicauda*) and its relatives belong to an ancient group that includes extinct giant water "scorpions" known as eurypterids. Their shared body plan – a fused head and thorax, a separate abdomen, and a shieldlike carapace – dates from the dawn of invertebrate evolution.

Stiff, tail-like telson is used as a rudder to help steer when swimming

**Carapace** is stiffened with a tough material called chitin, but it is not hardened with minerals like the brittle shell of a true crab

TOP VIEW

**Hinge** between the front part of the body (fused head and thorax) and the rear part (abdomen) helps the animal flex across the middle

A pincer (chela) on the end of each walking leg is used for grabbing prey and passing it forwards towards the mouth

Grinding plates called gnathobases at the base of the legs break up prey into pieces tiny enough to fit inside the small mouth

Males cling to females as they fertilize thousands of eggs that are then buried in sand

**Mating horseshoe crabs**
Atlantic horseshoe crabs (*Limulus polyphemus*) gather in huge numbers to breed along shallow sandy coastlines – much like their ancestors did millions of years ago.

## UNCHANGED BY TIME

The stability of some ocean habitats may help explain why evolutionary change in certain groups seems minimal. *Lingula* brachiopods (shelled animals that burrow in sediment) could be the least altered of all animals. Fossil shells from the Cambrian Period, 540 million years ago, are virtually identical to shells of modern-day *Lingula*.

*LINGULA* BRACHIOPODS

**Seeing salt**
The Dead Sea, an inland sea between Israel and Jordan, has an average salinity of about 34 per cent. The high rate of evaporation leads to the fomation of visible deposits of salt.

# salt water

The first oceans that formed on Earth some four billion years ago contained little salt, but were weakly acidic as a result of gasses released by volcanic activity. But today, as a result of dissolution of minerals from the land and the seafloor, the oceans contain over five trillion tonnes of salts from nearly 100 different elements. The concentration of salts in the oceans is between 3.3 and 3.7 per cent. Salinity is greater in warm, semi-enclosed seas such as the Mediterranean and the Red Sea. The two main components of ocean salts are sodium and chlorine, with smaller amounts of magnesium, sulphur, calcium, and potassium. If the oceans evaporated completely, salt residues would be equivalent to a layer 45 m (150 ft) thick over the entire planet.

## THE SALT CYCLE

In the chemical balancing act that maintains the salinity of the oceans, elements from mid-ocean ridges, volcanoes, and dissolved in rivers are balanced by the removal of salts by living organisms and the deposition of their remains on the seafloor. Some salts are also removed by chemical changes (mineralization) and by incorporation of sediments into the substrate as a result of shifts in the seafloor.

Clouds of volcanic ash

Spread of volcanic ash into rain clouds

Dust blown off land

Volcanic ash falls into the sea

Rivers carry minerals to the sea

Rain washes volcanic dust and gases into the sea

Minerals from seafloor incorporated into landmass

Uptake of salts by marine organisms

Deposition of salts from marine organisms

Minerals released by volcanic activity

**Eyes on stalks** allow the crab to see signals and threats from further away

**Waving claw**, present only in males and used for attracting a female, threat displays, and fighting; can be on left or right side

**Showing off**

Male crabs that have larger claws and an impressive waving display are more likely to attract a mate. It takes a strong male North American red-jointed fiddler crab (*Minuca minax*) to grow and wield a heavy appendage, but some males appear to gamble by growing impressive-looking lightweight claws that are actually too flimsy for fighting.

**Smaller claw** of male is used for feeding

**Walking legs** lack pincers

**CRUSTACEAN APPENDAGES**

Crabs, shrimps, lobsters, and crayfish are classified as decapod crustaceans, which all have a segmented body and one pair of appendages per segment. Decapod means "ten legs", referring to four pairs of walking legs and one pair that bears the claws. Other appendages serve as antennae, for collecting sensory information, and mouthparts. Crabs differ from other decapods in carrying their abdomen tucked under their body rather than extended behind them.

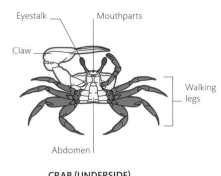

CRAB (UNDERSIDE)

# waving crabs

Many species of crab exhibit pronounced differences between males and females, a phenomenon known as sexual dimorphism. The males are often equipped with outsized claws (chelae), used for fighting and social signalling, including threat displays and courtship. Most fiddler crabs live on flat expanses, such as sandy beaches and mudflats, where visual signals can be seen from relatively long distances. Males dig breeding burrows and wave from the entrance to attract passing females. Before mating, the male plugs the burrow from inside to keep rivals out and protect the female and her eggs. She emerges when her offspring are ready to hatch into the sea.

**Female advantage**
The female fiddler crab uses both claws for feeding, unlike the male, which uses just one. The female is therefore able to pick up food twice as fast.

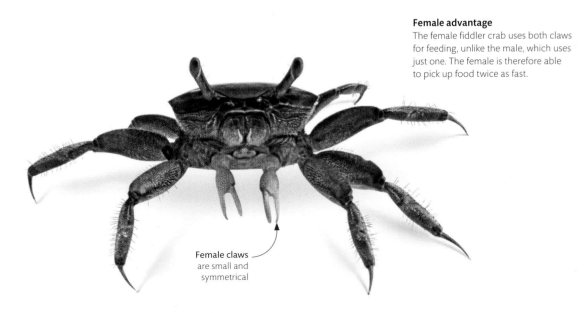

Female claws
are small and
symmetrical

# hunting above water

In bankside waters, including mangrove swamps, many predatory fish supplement underwater prey with targets on overhanging vegetation above the surface. Some leap vertically upwards to catch an insect, but archerfish also use a second tactic: they spit a jet of water to topple the target. They are remarkably accurate – even in rippling, cloudy waters. Their lightning-fast speed means they can also grab the prize before nearby competitors even see it fall.

Archerfish can leap up to twice their 30 cm (12 in) body length above water

**Jumping for prey**
Archerfish can strike prey on low-hanging leaves by launching their whole body out of the water. This energetic hunting method can often be more successful than shooting prey.

**Large eyes** provide acute vision, even in the low light of densely shaded mangrove swamps

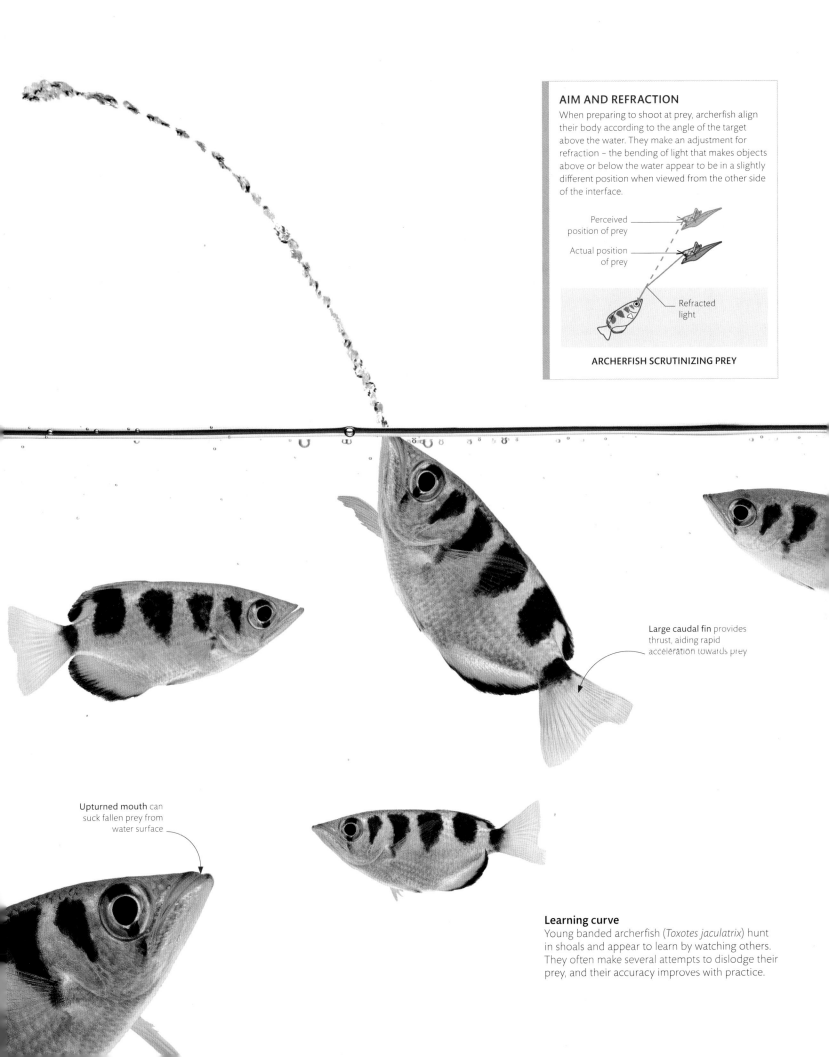

**Large caudal fin** provides thrust, aiding rapid acceleration towards prey

**Upturned mouth** can suck fallen prey from water surface

## Learning curve
Young banded archerfish (*Toxotes jaculatrix*) hunt in shoals and appear to learn by watching others. They often make several attempts to dislodge their prey, and their accuracy improves with practice.

**Étretat. The Cliff of Aval (1890)**
A Normandy seascape by Eugène Boudin features his characteristic luminous skies and tranquil waters, but also the striking arch of rock made famous by his pupil Claude Monet.

**Impression, Sunrise (1872)**
Claude Monet's hazy view of the port of Le Havre, in Normandy, is credited with giving rise to the name of the Impressionist movement. Awash with colours against a vague backdrop of ships and buildings, the work was vilified when it was exhibited in Paris in 1874.

*the ocean in art*

# impressions of the sea

Claude Monet is often regarded as the father of the Impressionist movement. In his later years, he paid tribute to the artist who taught him how to paint the sea. In 1858, Monet was an 18-year-old caricaturist from Normandy, France, when he first met Eugène Boudin, 20 years his senior. Boudin introduced the young Monet to the naturalistic approach of painting "en plein air" (outdoors), where he learned to absorb and capture the atmosphere of the coast.

Boudin was a master of expansive seascapes, skyscapes, and tranquil scenes of gentlefolk picnicking and promenading along the shore. Monet took on board his friend's advice that he should "learn to appreciate the sea, the light (and) the blue sky", but he went on to develop his own Impressionist techniques to portray the dramatic Normandy coastline. His focus was on capturing the colours of shadows in different lights, eliminating black to create a more vibrant effect. He painted the arch of rock called the Manneporte at Étretat up to 18 times, using quick dabs from loaded brushes to fix a ray of light or a cloud, and layering different tints while still wet instead of mixing colours.

In his journals, the French writer Guy de Maupassant describes Monet on the beach in Étretat in 1886, switching between five or six canvases as the light changed throughout the day. On one occasion, writes Maupassant, "he took a downpour beating on the sea in his hands and dashed it on the canvas – and indeed it was the rain that he had thus painted …"

“ A landscape is only an impression, instantaneous, hence the label they've given us. ”

CLAUDE MONET, *INTERVIEW FOR LA REVUE ILLUSTRÉE*, 1889

## CONFORMERS AND REGULATORS

Many marine animals – such as jellyfish, crustaceans, and starfish – have salt levels in their bodies that change, or conform, with their surroundings. Other animals, including most vertebrates, keep their salt levels constant: they regulate. Both conformers and regulators can survive the changing salinity in an estuary. The conformers tolerate extremes of salt inside the body; the regulators resist this from happening.

FRESH WATER · ESTUARY · OPEN OCEAN

Relative salt concentration in blood

Red dots indicate the limits of salt tolerance (beyond which the animal dies)

Salt levels inside a conformer

Salt levels inside a regulator

Salt (parts per thousand) in surrounding water

0   5   10   15   20   25   30   35   40

**SALINITY IN CONFORMERS AND REGULATORS**

# adapting to changing salinity

Salt levels in the open ocean are relatively stable, at around 35 parts per thousand. The influence of fresh water added by rivers has little effect, except near the coasts. Most ocean organisms – just like those in freshwater habitats – are so adapted to the water around them that they cannot survive changes in salinity: they are described as stenohaline. In estuaries, where salt levels fall and rise with the tides, life is euryhaline: it tolerates fluctuating salinity.

**Tears** produced by lachrymal glands contain a high concentration of salt; the glands secrete excess salt when the turtle gets dehydrated

**Horny beak** is used to crush prey such as fish and marsh snails

**Foot** pushes backwards against the water for thrust

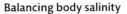

**Skin** is cornified (thickened) with plenty of reinforcing protein called keratin, which helps to reduce absorption of excess salt or water

### Balancing body salinity
The only claw-footed turtle to thrive in estuaries, the diamondback terrapin (*Malaclemys terrapin*) has tough skin to resist the variable salinity and tear glands to secrete excess salt. If its salt levels rise, it spends more time in fresh water, avoids the saltiest prey, and lifts its head to drink directly from the fresh water of falling rain.

### Clawed, webbed feet
Like freshwater turtles and terrapins, the diamondback has clawed feet, rather than flippers, for walking on the ground or estuary sediment. Webbing between the toes helps with propulsion in the water.

# coral reefs

**Among the most complex** and beautiful of Earth's ecosystems, reefs are built by colonial animals. These massive structures are home to diverse, colourful communities of organisms, many of which have become highly specialized to find their own niche.

## Bottle colony
The filter-feeding "bottles" of an Indo-Pacific blue sponge (*Lamellodysidea chlorea*) are connected in a sprawling colony. In coastal seas murky with sediment, *Lamellodysidea* sponges can smother living corals ▪ and may come to dominate a reef.

*Didemnum* sea squirt grows as a brightly coloured, filter-feeding colony – either vivid green or pink – on almost any available surface, including the bodies of sponges

An opening, called an osculum, at the top of each bottle-shaped chamber releases water that has been filtered of food

The chamber walls are peppered with cells that carry pores; each pore, called an ostium, allows water to enter the sponge

# simple bodies

Oceans are home to the simplest animals – sponges. Some animals in this group can grow into giants more than 1 m (3 ft) across, but in all of them – no matter what the size – their constituent cells are so loosely connected that if the entire body is fragmented, each piece may retain its ability to develop into a new individual. Nevertheless, just as in other animals, the cells cooperate to keep the entire structure alive. They help to circulate water through the porous sponge to extract suspended food.

Fragile skeleton grows at great depths

**GLASS SPONGE**

Barrel sponge, the largest type of sponge, can reach 1.8 m (6 ft) in diameter

**DEMOSPONGE**

*Clathrina* grows as tangled tubes

**CALCAREOUS SPONGE**

### Types of sponge
A sponge's supporting skeleton is made of different possible materials: fibres of collagen (a protein) or hard spicules (mineral needles). Demosponges, generally the softest sponges, are typically dominated by collagen, while glass and calcareous sponges are supported, respectively, by silica or calcite spicules.

### BODY ORGANIZATION
The different kinds of cell that make up the body of a typical sponge are organized around porous, bottlelike chambers. Collar-flagellate cells in the body lining have beating hairs that maintain water flow: in through the walls and out through an opening at the top. Each hair sits within a permeable collar that traps particles of food carried in the water stream.

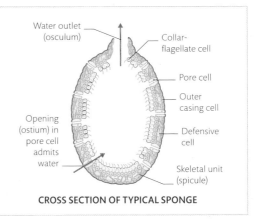

Water outlet (osculum)

Collar-flagellate cell

Pore cell

Outer casing cell

Opening (ostium) in pore cell admits water

Defensive cell

Skeletal unit (spicule)

**CROSS SECTION OF TYPICAL SPONGE**

Each polyp carries a ring of tentacles for catching plankton; polyps at the ends of coral branches have six tentacles, those on the sides have twelve

Epidermis (surface skin) of the coral contains brown-pigmented algae called zooxanthellae, which supplement the coral's nutrition through photosynthesis

Each branch of the coral is supported by a rocky core that makes up most of its bulk

# making rock

Some animals affect their surroundings more profoundly than others. Many corals use minerals from sea water to construct rocky skeletons that accumulate over hundreds – or even thousands – of years, producing the most celebrated of all ocean features: the coral reef. The living coral persists as scarcely more than a thin veil with tiny plankton-catching tentacles. But the rocky foundation it has made can be a kilometre (half a mile) thick and stretch for hundreds of kilometres across the seabed.

**Coral branches**
The branches of a stony coral (*Acropora* sp.) colony consist of a hard, rocky core clad in a "skin" that carries tentacled polyps. Nutrients harvested from trapped plankton are shared around the polyp colony by the skin.

**A branching skeleton**
has a large enough surface area to support many thousands of polyps in life

**Dead coral**
Stripped of living tissue, this white coral skeleton is virtually pure calcium carbonate – a chalky mineral made by the coral's living layer.

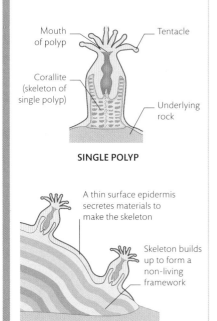

## BUILDING A REEF
Microscopic planktonic coral larvae settle onto rock, where they develop into polyps. Each polyp, no bigger than a grain of rice, builds a tiny, cup-shaped rocky skeleton, known as a corallite. Over time, more interconnected polyps are generated as the colony expands outwards on the surface, while the skeleton beneath thickens – by about 0.5 cm (¼ in) per year – to become the rocky foundation of a reef.

Mouth of polyp

Tentacle

Corallite (skeleton of single polyp)

Underlying rock

**SINGLE POLYP**

A thin surface epidermis secretes materials to make the skeleton

Skeleton builds up to form a non-living framework

**CORAL REEF**

# synchronizing spawn

Sexual reproduction for most types of ocean animals involves broadcasting sex cells into open water and relying on chance that fertilization will happen. For corals, separate patches of the same species must synchronize their spawning so that sperm and eggs are released at the same time. Environmental cues – such as a seasonal rise in temperature or phases of the Moon – trigger one of the most spectacular reproductive events on the planet: when a reef bursts with clouds of coral spawn. It has been estimated that a million eggs come from each square metre of reef.

**Gamete bundle** emerges from mouth of polyp

**Reproductive polyp**
Most corals are hermaphrodites: the polyps in their colony eject round gamete bundles that break apart into sperm and eggs within about half an hour of their release.

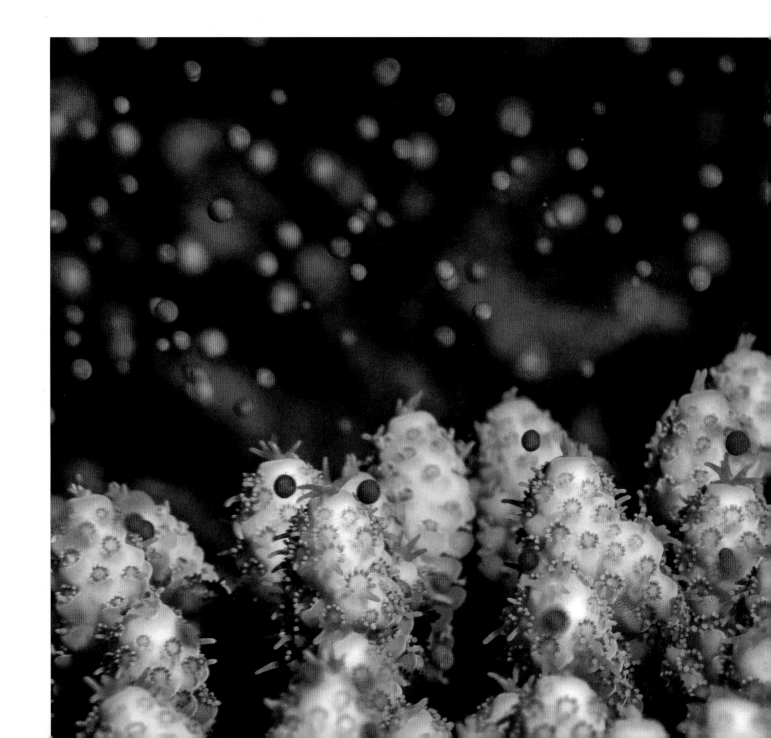

## CORAL SEXUAL LIFE CYCLE

Corals and related anemones have a life cycle dominated by the sedentary polyp: a flask-shaped animal with stinging tentacles for catching planktonic prey. Colonies of coral polyps release their gamete bundles in synchrony to maximize the chances of fertilization. Each fertilized egg develops into a tiny, flat-bodied planula larva, which mixes with the drifting plankton. If it survives the attention of predators, the larva eventually settles on the seabed and transforms into a polyp. It then multiplies asexually to form a new coral colony.

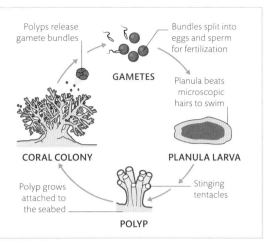

Polyps release gamete bundles

Bundles split into eggs and sperm for fertilization

**GAMETES**

Planula beats microscopic hairs to swim

**CORAL COLONY**

**PLANULA LARVA**

Polyp grows attached to the seabed

Stinging tentacles

**POLYP**

### Releasing gametes

In some reefs around the world, all corals spawn simultaneously, but in the Red Sea, species vary in their spawning season. This species of *Acropora* coral spawns in early summer on the night of a full Moon, releasing pink bundles of gametes by the billion.

**Epithelium (surface skin)** secretes the gorgonin, which forms the solid inner core that supports the colony

**The base of each polyp** connects with neighbouring polyps through channels called solenia, which share nutrients through the colony (see p.132)

**Eight-tentacled polyps** illustrate the eight-fold symmetry that is characteristic of sea fans and other types of octocoral

**Skin colour** varies between colonies of this species: some are yellow instead of red

**Dichotomous branching,** by which a branch repeatedly splits in two as it grows, produces a fan shape that is effective at capturing plankton over a wide area

**Growing across the current**
Sea fans typically grow in one plane that is perpendicular to the direction of the current, ensuring that their fan shape is in the best position for trapping drifting plankton.

# horny skeleton

The best-known corals are those that lay down the huge rocky foundations of a reef. But related animals build colonies in a different way. Sea fans grow branches supported by a core of hornlike protein called gorgonin. They stretch upwards from the seafloor or horizontally as they suspend over steep drop-offs on the outer edge of a reef. Here, far below the ocean surface, they use their tiny polyps to trap plankton.

**Branches of the deep**
A single branch of the colourful sea rod (*Diodogorgia nodulifera*), from the Caribbean, is covered with plankton-trapping polyps. Unlike many stony corals growing near the surface, this sea fan contains no light-absorbing, photosynthesizing algae, so it is able to grow in the dark depths.

**VARIED SUPPORT**
Sea fans belong to a group called octocorals, so named because their polyps carry eight tentacles. Support structures in octocorals are more diverse than in the stony, six-tentacled hexacorals. Some, the soft corals (see pp.132–33), lack any hard skeleton at all. Gorgonians, which include sea fans, are supported by horny material called gorgonin, while the organ-pipe coral (*Tubipora musica*) is reinforced by chalky minerals.

**STRUCTURE OF ORGAN-PIPE CORAL**

coral reefs

Polyps can retract if danger threatens, or extend their tentacles when catching plankton

Polyps containing **photosynthetic algae** cover the "mushroom's" cap, where they are exposed to light

**Soft-bodied coral**
Soft corals lack the hard, rocky skeletons of reef corals. Some, such as the rough leather coral (*Sarcophyton glaucum*), form giant, fleshy, mushroom-shaped colonies.

# living in a colony

Colonial animals are common in the ocean. A mature coral consists of thousands of tiny, anemonelike polyps. Each polyp uses its stinging tentacles to catch plankton but is linked to its neighbours so that nourishment is shared. All the polyps develop from the same fertilized egg and are genetically identical, and the entire colony functions like a single superorganism. By spreading over a wide area, the colony can maximize its catch of food as well as the number of eggs it is able to release during spawning.

**Fleshy colony**
Dozens of polyps pack just a thumb-sized area of a *Sarcophyton* leather coral. As well as catching plankton, each polyp is packed with photosynthetic algae, from which they gain extra nutrients. The polyps project from a mass of fleshy tissue called coenenchyme.

**SHARING NUTRITION**

Each coral polyp has a gut cavity that ingests planktonic food and ejects undigested waste. Nutrients from digested prey are spread throughout the colony. In stony corals, this occurs in the thin "skin" overlying the rocky skeleton. In soft corals, canals called solenia penetrate deeper into the fleshy colony to share the food.

**TYPICAL SOFT CORAL STRUCTURE**

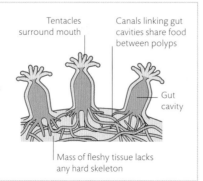

Tentacles surround mouth

Canals linking gut cavities share food between polyps

Gut cavity

Mass of fleshy tissue lacks any hard skeleton

## Stony corals

Also known as hard corals, these are the architects of the world's coral reefs. True stony corals have rigid skeletons made of calcium carbonate – in the form of a mineral called aragonite – and each polyp sits in its own protective cup, called a corallite. In areas with strong waves these colonies develop into robust mounds or flattened shapes; in sheltered seas, the same species can form more intricate shapes with delicate branching patterns.

**Dense groups,** or thickets, up to 2.4 m (8 ft) across and 1.2 m (4 ft) high form in shallow waters

**Large, thick branches** resembling elk antlers provide shelter for many reef species

**STAGHORN CORAL**
*Acropora cervicornis*

**ELKHORN CORAL**
*Acropora palmata*

## Soft corals

Also known as octocorals because their polyps have eight tentacles, these corals are soft and bendable. Often treelike in appearance, these non-reef-building colonies are supported by hornlike cores and protected by fleshy outer rinds.

**Treelike species** found in deep waters of the North Atlantic and the Barents Sea

**Stiff, bushlike** colony grows in a single plane

**CARNATION CORAL**
*Gersemia fruticosa*

**GORGONIAN SEA FAN**
*Subergorgia*

**Flexible, fleshy branches** enable coral to tolerate buffeting by strong ocean currents

**Eight-tentacled polyp** captures passing plankton to obtain nutrients

## Fleshy and flexible

Commonly known as the Kenyan tree coral (*Capnella imbricata*), this branched soft coral emerges from a single "trunk" that is fixed to a rock. The colonies inhabit coastal shores and rocky reef slopes and can grow to around 45 cm (18 in).

Large polyps branch in many directions, and tentacles extend to catch passing zooplankton

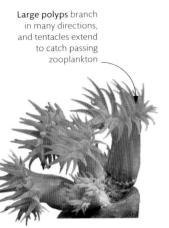

**TUBE CORAL**
*Tubastraea coccinea*

Vertical, cylindrical columns up to 3 m (10 ft) tall grow from an encrusted base

**PILLAR CORAL**
*Dendrogyra cylindrus*

Long, meandering valleys in dome each contain several polyps

**BOULDER BRAIN CORAL**
*Colpophyllia natans*

Horizontal plates are formed of tiny branchlets that grow vertically and horizontally

**TABLE CORAL**
*Acropora cytherea*

Red branches are edged with white polyps and move with ocean currents

**GORGONIAN SEA WHIP**
*Ellisella ceratophyta*

Base, or peduncle, anchors in sand or mud in shallow waters

**SEA PEN CORAL**
*Pteroeides*

Open polyps ready to consume prey; they close before eating again

**MUSHROOM CORAL**
*Anthomastus*

Fingerlike "branches" are covered with retractable polyps

**SPAGHETTI FINGER LEATHER CORAL**
*Sinularia flexibilis*

# corals

Corals are found in all the oceans of the world, but reef-building species are found mainly in shallow tropical and subtropical seas. Coral is formed from huge colonies of individuals called polyps that feed on plankton. Colonies that live in shallow, warmer waters host zooxanthellae, which use carbon dioxide from the coral's respiration in photosynthesis and, in turn, pass nutrients back to the coral.

Green concentrations of
**zooxanthellae** are found in
the polyp's column, which
contains the gut cavity

**Algae in the flesh**
When a single coral polyp is
magnified, the zooxanthellae are
visible as spots of green-coloured
cells. Each cell has light-absorbing
chlorophyll – the same pigment
found in photosynthesizing plants.

**Transparent,
jelly-packed tissues**
help to transmit light
to zooxanthellae in
the gut lining

# powered by sunlight

Animal flesh is nourished by food that has been digested and assimilated,
but some types of marine animal are partly solar powered. Many corals
and some hydrozoans, anemones, sponges, and molluscs contain
photosynthetic algae called zooxanthellae. These grow and multiply in the
animal's sunlit tissues and use carbon dioxide from its respiration to make
food, much of which passes to the host and is key to healthy growth.

**Skeleton from photosynthesis**
*Millepora* fire hydrocorals – named for their painful sting
– are colonial hydrozoans with hard skeletons, like those
of true stony corals. Their dependence upon symbiotic
zooxanthellae restricts them to brightly lit shallow water.
They use much of the food gained from the algae's
photosynthesis to build their chalky skeletons.

## ZOOXANTHELLAE

The zooxanthellae of marine animals belong to a
group of algae known as dinoflagellates. These have
free-living forms that swim with whiplike structures
called flagella, but when swallowed by coral polyps,
they lose their flagella and become cysts in the
polyp's gut lining. A natural yet vulnerable
balance exists between the swimming and
encysted forms. Warming of the sea is leading
to a net loss of cysts, bleaching corals and often
killing them.

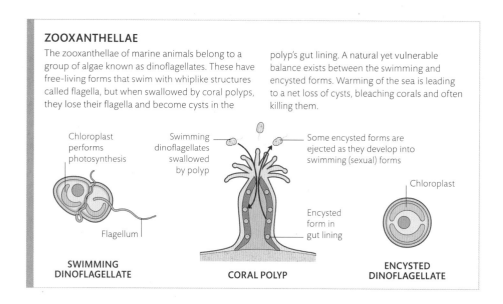

Chloroplast
performs
photosynthesis

Flagellum

**SWIMMING
DINOFLAGELLATE**

Swimming
dinoflagellates
swallowed
by polyp

Some encysted forms are
ejected as they develop into
swimming (sexual) forms

Encysted
form in
gut lining

**CORAL POLYP**

Chloroplast

**ENCYSTED
DINOFLAGELLATE**

**The longer tentacles** of many anemone species can reach further to overcome large prey

**Solar-powered mat**
Like some other anemones, Haddon's carpet anemone (*Stichodactyla haddoni*), of the Indo-Pacific, supplements nutrients from its prey with food produced by algae growing in the flesh of its wide oral disc. The algae use energy from sunlight penetrating the water to photosynthesize and make sugar, which they share with the anemone.

**Stinging tentacles**
Anemones vary in the potency of their venom. Most trap tiny planktonic animals, but some, such as this dahlia anemone (*Urticina felina*) are able to immobilize bigger prey, including urchins.

**Knoblike ending** of each short tentacle is coated in nematocysts (stinging cells)

# stinging for prey

For an animal that is attached to the seabed, an anemone is a surprisingly efficient predator: a fleshy column, or polyp, crowned with stinging tentacles reaches up into the water to catch passing animal prey. The tentacles bring the paralysed victim to the mouth at the centre of the polyp, which swallows the meal into the digestive cavity below. Carpet anemones are among the largest polyps of a tropical reef: a single oral disc, covered in short, budlike tentacles, can span nearly 1 m (3 ft).

### FIRING VENOM
Anemones, along with corals and jellyfish, are cnidarians. These soft-bodied predators owe their stinging capabilities to unique cells, called nematocysts, in their skin. Each nematocyst contains a capsule of venom and a barbed thread like a miniature harpoon. Contact with prey triggers the harpoon to fire, penetrating the victim with a chemical cocktail that can paralyse muscles.

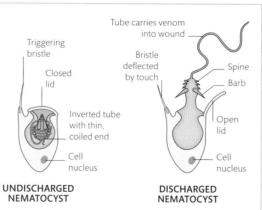

Triggering bristle

Closed lid

Inverted tube with thin, coiled end

Cell nucleus

**UNDISCHARGED NEMATOCYST**

Tube carries venom into wound

Bristle deflected by touch

Spine

Barb

Open lid

Cell nucleus

**DISCHARGED NEMATOCYST**

**Fleshy oral disc** surrounds a central mouth and contains contracting muscle fibres that raise its carpet of stubby tentacles upwards into the water – and closer to prey

**Muscular column** contains a gut cavity that digests prey and absorbs nutrients

# coral reefs

Reef building corals live in shallow tropical waters, where the sea-surface temperature ranges from 20° to 35°C (68° to 95°F). They need a rocky surface on which to build, normal marine salinity, and waters that are free from excess sediment. Vast populations of tiny coral polyps – individuals are often no larger than a pea – work together with a host of other organisms to construct and maintain the largest living structures on Earth. The Great Barrier Reef off northeastern Australia, which comprises over 3,000 separate reefs, is even visible from space. Coral reefs are highly sensitive to changes in their ocean environment and are gradually dying as a result of rising temperatures, ocean acidification, and reduced oxygen levels.

**Rich pickings**
As dusk falls, shoals of epaulette soldierfish (*Myripristis kuntee*) emerge from the cover of a reef in Palau. Their big eyes help them hunt through the night on planktonic animals, such as crab larvae.

## CORAL ATOLL FORMATION

In the open ocean, coral atolls originate where reef organisms have constructed a fringing reef around the shores of newly formed volcanic islands. When volcanic activity ceases, the island subsides due to a combination of plate movement and weathering. The reef on its flanks continues to grow upwards, first becoming a barrier reef – separated from the land by a lagoon. Later the island disappears, resulting in the formation of an atoll – a ring of coral islands with no central landmass.

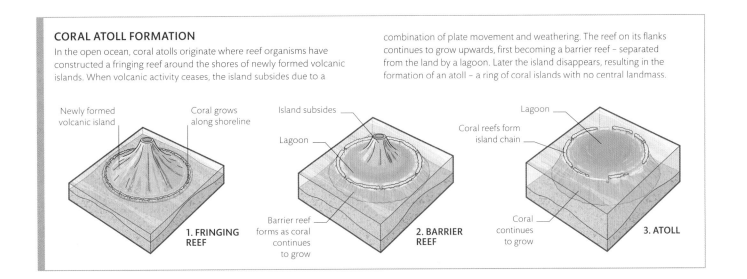

Newly formed volcanic island

Coral grows along shoreline

**1. FRINGING REEF**

Island subsides

Lagoon

Barrier reef forms as coral continues to grow

**2. BARRIER REEF**

Lagoon

Coral reefs form island chain

Coral continues to grow

**3. ATOLL**

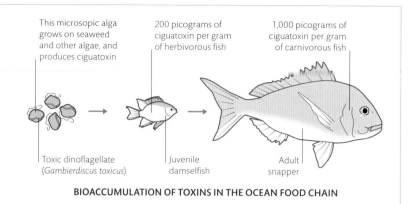

**PASSING ON POISON**

Some animals that eat zoanthids, such as crustaceans, seem to tolerate their poison. Many ocean toxins get passed along food chains in this way. Ciguatera syndrome in humans is caused by eating fish that contains high levels of ciguatoxin – a poison produced by algae called dinoflagellates. This poison is not excreted after the algae are consumed, so its concentration rises to potentially lethal levels in consumers higher up in the food chain – including fish served up on diners' plates.

This microsopic alga grows on seaweed and other algae, and produces ciguatoxin

200 picograms of ciguatoxin per gram of herbivorous fish

1,000 picograms of ciguatoxin per gram of carnivorous fish

Toxic dinoflagellate (*Gambierdiscus toxicus*)

Juvenile damselfish

Adult snapper

**BIOACCUMULATION OF TOXINS IN THE OCEAN FOOD CHAIN**

# blooming with poison

Tropical sunlit coastlines around the world can be gaudy with colourful zoanthids. Like anemones, these animals produce large, soft polyps. Like corals, they grow in colonies – although they lack a hard, reef-forming skeleton. The biggest zoanthid polyps resemble mesembryanthemum flowers, but their striking colours may serve as a warning of their toxicity. Many zoanthids harbour bitter chemicals called alkaloids that are poisonous and deter grazing animals. At least some of these poisons may be derived from planktonic organisms trapped by the zoanthid's tentacles.

**Blue base of the tentacles** carries a different form of the alkaloid pigment compared with that in the green tips

**Lethal pigment**
A zoanthid obtains both its colour and its poison from alkaloid chemicals known as zoanthoxanthins. The alkaloids produce the vivid blue and green of the tentacles, but they are also toxic to nerves and muscles in other animals. Also present – but hidden by the alkaloids – are photosynthetic algae that provide the zoanthid with extra food.

**Tentacles** may fluoresce due to the colourful alkaloid: some of the light absorbed by the chemical during the day is emitted at night, producing a greenish glow

**Radioles (side-branches)** of each palp have beating cilia (microscopic hairs) that waft particles of food towards the mouth

**The operculum** is a modified, disc-shaped palp that is used to plug the entrance to the worm's tube for protection when the animal withdraws inside

**The calcareous tube** is strengthened by a triangular ridge running along the top

### Coral-living worm

The palps of a star fan worm (*Pomatostegus stellatus*) spread out in the shape of a horseshoe. They are packed with oxygen-absorbing pigment and lined with food-collecting hairs. The rest of the worm lives inside a calcareous (chalky) tube buried within the rocky skeleton of a tropical stony coral.

### Exposed tube

The hard tube of *Spirobranchus lamarcki*, a European relative of the star fan worm, is visible – typically attached to a stone, rock, or occasionally the carapace of a crab.

**Pink hue of palps** comes from blood pigments in their tissues; the pigments chemically bind to oxygen and extract it from water for respiration

# fanning the water

Of the thousands of species of marine annelid (segmented) worms, many use their muscular bodies for burrowing in sediment – just like earthworms, their cousins on land. Other marine annelids can swim through open water. But some, called fan worms, live in tubes, compromising their ability to swim or burrow away from danger. A fan worm's front end reaches out into the water, where a crown of fingerlike structures, called palps, sweeps the water for oxygen and particles of food. When danger threatens, the body uses retractor muscles to pull the exposed parts completely inside the tube.

### TYPES OF TUBE

Fan worms called serpulids, such as the star fan worm, build hard, chalky tubes made from calcium carbonate secreted by glands in the collar around the base of their palps. Other types of fan worm – the closely related sabellids – build soft, membranous tubes made from mucus mixed with sediment collected by the palps.

Hard shell is fused to rock or stone on one side

Soft tube is partially buried in sand or mud

**SPIRORBIS** (A SERPULID)

**SABELLA** (A SABELLID)

**Water purifiers**
There are more than 10,000 species of marine bivalves. All are filter-feeders, able to remove contaminants from sea water; an oyster can filter 95 litres (25 gallons) a day.

*spotlight species*

# giant clams

Giant clams (*Tridacna*) are the biggest non-colonial animals that grow fixed to the coral reef. These heavyweight molluscs bask in the tropical sunlight of the Indian and Pacific oceans from east Africa to the Pitcairn Islands.

Like other bivalve molluscs, a giant clam has a shell formed of two valves, connected by a hinge. It contains zooxanthellae – a type of algae – that photosynthesize and make 90 per cent of the clam's food. The remainder of its nutrition comes from filter-feeding plankton. In some species, the fleshy mantle is vividly coloured with pigments or reflective crystals, which help to screen ultra-violet and other potentially harmful radiation.

Just like food-making algae in coral, the giant clam has a mutually beneficial relationship with the zooxanthellae as the clam passes carbon dioxide and nitrogen compounds that are needed to make nutrients into the zooxanthellae. This symbiosis is productive: a record-breaking specimen of *Tridacna gigas*, the largest species, weighed half a tonne.

Like so many other iconic animals on the reef, giant clams help to engineer the complex community around them. Over time, their shells are colonized with life, while many animals – from crustaceans to fish – live over the mantle, some as parasites. Zooxanthellae occasionally expelled from the mantle may provide food for plankton-eaters too.

Giant clams are hermaphrodites, but release most of their sperm before their eggs to reduce the chance of self-fertilization. Like most other molluscs, the fertilized eggs develop into tiny larvae that swim in the plankton, before metamorphosing into adults and settling on the ocean bottom.

**Patterned giants**
The smooth giant clam (*Tridacna derasa*) of barrier reefs and island atolls, is the second largest species, growing to 60 cm (24 in) long, and has a particularly brilliant mantle pattern of blue or green.

The cloud of ink contains chemical irritants that help to repel predators

**Defensive ink**
When disturbed, this sea hare – a kind of sea slug – releases a cloud of acrid purple ink. The ink is partly derived from the sea hare's diet of algae.

**Poison alert**
The vivid colours of a variable neon sea slug (*Nembrotha kubaryana*) are aposematic, meaning that they act as a warning signal to predators. This Indo-Pacific species has toxic slime, which comes from eating a particular type of sea squirt – a soft-bodied, filter-feeding animal of the seafloor.

# collecting weaponry

Soft-bodied molluscs lacking hard shells need alternative forms of protection against predators. Many sea slugs equip themselves with defences acquired from their food. Some graze on toxic organisms and harbour the poisons to ooze their own noxious slime or eject repelling ink. Others eat the stinging tentacles of jellyfish or anemones and preserve the stingers (nematocysts), which become incorporated into their bodies yet remain fully capable of firing.

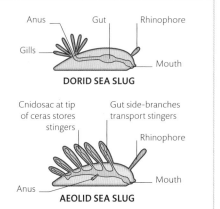

**STORING STINGERS**
The variable neon sea slug belongs to a group of sea slugs called dorids, which have a plume of gills and rely on toxic slime. Another group, the aeolids, store stingers from their prey in each fingerlike projection, or ceras (plural: cerata), on their back. The microscopic stingers ingested with their food are transported through side-branches of the gut and held in the cerata tips, ready to fire when a predator attacks.

Anus   Gut   Rhinophore
Gills
Mouth
**DORID SEA SLUG**

Cnidosac at tip of ceras stores stingers   Gut side-branches transport stingers
Rhinophore
Anus   Mouth
**AEOLID SEA SLUG**

**Orange markings** help reinforce the colourful warning of the animal's toxicity, but some equally poisonous individuals of this variable species are only green and black

**Tentaclelike sense organ**, called a rhinophore, tastes the water and can detect sea squirt prey

**Plumelike gills**, arranged in a circle, absorb oxygen from the surrounding water

### Quick change
Octopuses that are more active in sunlight than at night are the best colour-changers, and the aptly named day octopus (*Octopus cyanea*) is the champion, being recorded switching 1,000 times over a seven-hour period. It uses black stripes when courting and turns dark when hunting.

## PIGMENT CONTROL
Colour changing in many animals – including octopuses and other cephalopods – is achieved by specialized skin cells called chromatophores. Each chromatophore contains a pigment, the colour of which is revealed when it disperses from a central point. Cephalopod chromatophores are especially complex, being controlled by tiny muscles that dilate a pigment-containing sac to spread the colour – all triggered by nerve impulses voluntarily fired from the brain.

Muscle cells are relaxed
Pigment granules

Muscle cells contract
Sac dilates

Granules spread out

**PALER SKIN COLOUR**

**DARKER SKIN COLOUR**

**Each arm** carries two rows of stalked suckers that can grip prey and other objects

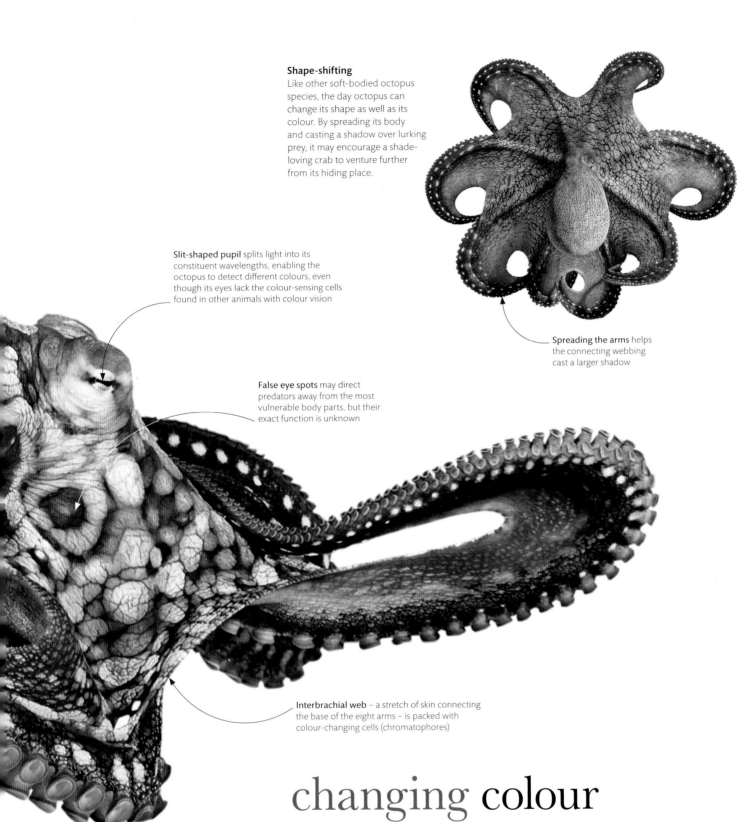

**Shape-shifting**
Like other soft-bodied octopus species, the day octopus can change its shape as well as its colour. By spreading its body and casting a shadow over lurking prey, it may encourage a shade-loving crab to venture further from its hiding place.

**Slit-shaped pupil** splits light into its constituent wavelengths, enabling the octopus to detect different colours, even though its eyes lack the colour-sensing cells found in other animals with colour vision

**Spreading the arms** helps the connecting webbing cast a larger shadow

**False eye spots** may direct predators away from the most vulnerable body parts, but their exact function is unknown

**Interbrachial web** – a stretch of skin connecting the base of the eight arms – is packed with colour-changing cells (chromatophores)

# changing colour

Where sunshine penetrates the sea, many animals flaunt colours, and some even change colour when their behaviour alters. The switch can occur rapidly because it is controlled by lightning-fast nerve impulses. Cephalopods, such as octopuses and cuttlefish, are among the most impressive colour-changers. Within seconds, an octopus sensing danger can blend in with the seabed and vanish, or, spotting a potential mate, impress with bold patterns.

*the ocean in art*

# art meets science

Haeckel, Kunstformen der Natur.                    Tafel 43 — *Aeolis.*

Nudibranchia. — Nacktkiemen-Schnecken.

Natural history illustration came of age in the 19th century in response to public fascination with voyages to formerly remote regions and a scientific need for accurate representation of specimens. The sea was a new frontier for exploration, and the complexity and vibrancy of its myriad life-forms was both a challenge and an inspiration to the artists recording them.

Describing sea creatures was historically challenging as specimens rapidly lost their colours and shrank after being preserved in alcohol. The Zoological Society of London's solution was to use art painted in situ. From 1812, tea merchant John Reeves spent 19 years in Macao, where he commissioned Chinese artists to create detailed paintings of fish. Many of these paintings became iconotypes: the basis for the description of a new species.

**Nudibranchs (1899–1904)**
Crossing the boundaries between scientific illustration and art, Ernst Haeckel's nudibranchs (soft-bodied marine gastropods) are remarkable for their detail and harmonious arrangement.

Britain's HMS *Challenger* returned from a four-year voyage around the world in the 1870s, with samples trawled from many different ocean depths and new data on the chemistry of sea water, currents, sea temperatures, and the geology of the seafloor. One of many enthusiasts fired up by the findings was German doctor and zoologist Ernst Haeckel. His major work *Kunstformen der Natur* features 100 stunning lithographs that reveal the symmetry and beauty of the ocean's smallest life-forms.

A ground-breaking 1855 study of the Atlantic by US naval officer Matthew Fontaine Maury inspired naturalist James M. Sommerville to commission a dazzling watercolour illustration of its species.

**Ocean Life (1859)**
An idealized assembly of 75 vibrant sea creatures and corals from a variety of geographical zones was created by physician and naturalist James M. Sommerville to illustrate his pamphlet *Ocean Life*. Art professor Christian Schussele's watercolour for the lithograph has all the hallmarks of a still-life painting.

> ❝ The grouping may be considered, rather as an intellectual truth than as a literal view. It might be termed a mental oasis of the sea... ❞

JAMES M. SOMMERVILLE, *INTRODUCTION TO OCEAN LIFE*, 1859

# cleaning hosts

Sunlit coral reefs are fertile, physically complex places with a rich biodiversity. As a result, more species evolve to be specialized in what they do, reducing competition in such a crowded habitat, and many are driven into close relationships. Shrimps that once scavenged for food now pick parasites and dead skin off living hosts and even advertise their services at cleaning stations. In this mutually beneficial symbiosis, visiting fish "clients" are cleared of potentially harmful pests, while the cleaner shrimps get a meal.

**Cleaner shrimp** may enter the host's mouth to pick food from its jaws and teeth

**Cleaning the hunter**
Some cleaner shrimps form strong bonds with moray eels. By sharing the eel's lair, they get some protection from predators.

**Colourful cleaner**
On coral reefs in the Indo-Pacific, the distinctive red-and-white pattern of the scarlet skunk shrimp (*Lysmata amboinensis*) – named for its stripes – may serve as a visual signal to fish at cleaning stations. Several species of skunk shrimp act as efficient fish cleaners in tropical waters around the world.

**Long white antennae** may be waved around as part of the visual signal that advertises the shrimp's cleaning behaviour to a "client" fish

**Red stripes** are caused by a chemical pigment called astaxanthin in the crustacean's carapace

## CLEANER FISH

Fish from groups as diverse as gobies and wrasses have independently evolved cleaning behaviour on coral reefs. Like cleaner shrimps, they consume skin parasites, and most share a similar striped colour pattern, the stripes often being blue or dark. Their cleaning station is typically a conspicuous seabed feature, such as a coral outcrop. Here, "client" fish gather and may solicit the attentions of waiting cleaners through their posture or swimming movements.

Fig. 2.

**BLUESTREAK CLEANER WRASSE**
(*LABROIDES DIMIDATUS*)

**Large eyes** are used to spot "clients"; in dim light, shrimps may be more wary of approaching potentially dangerous predators and exaggerate their pre-cleaning signals, such as leg rocking and waving antennae

**Legs** rock the body from side to side when approaching a shrimp-eating "client", helping to reinforce the cleaning intent and avoid accidental attack by a predator

# making a mask

**Under cover**
The spider decorator crab (*Camposcia retusa*) is so devoted to staying hidden that the only parts of its body not covered by debris and living things are those it uses to collect the masking material – its pincers.

Although crabs can famously use their pincers in self-defence, many species adopt a less confrontational approach: they wear a disguise. By plastering their body with flotsam or cuttings – anything from scraps of seaweed to colonial animals like sponges – they can blend in with the seabed. Much of what they wear keeps on growing, becoming a living mask that improves the deception over time.

**Slender pincers** have sharp cutting edges for snipping pieces of algae or sponge

**Body debris** includes fragments of seaweed, sponges, small stones, and shells collected by the crab

**Hooked hairs** grow in bunches, which provides a stronger grip when holding debris

## Velcro crab

The body of the spider decorator crab is covered in hook-tipped hairs. The hooks work like Velcro, securing objects to the body transferred there by the pincers.

### CRAB DÉCOR

Some crab collectors are specialists. Those of the family Dromiidae camouflage themselves with sponges. They have hooked rear legs for transferring a sponge to the back of their shell. Pom-pom crabs (*Lybia*) even use décor with a sting: they carry tiny anemones (*Triactis*) in their pincers to ward off predators.

Anemones held by pincers

Sponge in place on crab's back

**POM-POM CRAB**

**SPONGE CRAB**

**The legs**, as well as the body, are covered in hairs, so practically the entire crab can carry debris

**Yellow or red sponges** stay alive and keep growing when attached to the crab

**Brightly coloured sponges** may produce noxious chemicals that repel any predators not fooled by the disguise

Fleshy projections (papillae) cover the body

**Anus** serves as both an exit for excreted waste and an inlet for oxygenated water required for respiration

**Sharing resources**

The thorny sea cucumber (*Colochirus quadrangularis*) lives in sunlit waters to about 115 m (380 ft) deep. When feeding, it rears upwards and spreads its tentacles to trap food suspended in the water. Some of the food may also be taken by tiny shrimps living in among the tentacles.

# seabed feeding

Animals that live sedentary or slow-moving lives on the seabed make the most of the food that comes to them. Suspension-feeders trap particles of living or dead material from the water, using bristles or tentacles. Detritivores specialize in dead material, known as detritus, which can be trapped in the water or gathered from the seabed. In deep water, detritus raining from above replaces sunlight as the main source of energy.

**Tentacles** carrying food are inserted one by one into the mouth

**Feeding tentacles**

The reef-dwelling sea apple (*Pseudocolochirus violaceus*) is a colourful sea cucumber that uses frilly tentacles to trap tiny particles, such as algae, zooplankton, and dead material, from sunlit shallow water.

**Body** is flexible and muscular

**Feathery feeding tentacles** trap particles of food

## WHALE FALL

Detritivores recycle matter and energy back into the food web. A dead whale first provides food for mobile scavengers, including sharks and hagfish, then slower ones like sea pigs (a type of sea cucumber), then colonists such as worms and crustaceans. When the flesh is gone, communities of bacteria in the bones produce hydrogen sulphide and use it in place of sunlight to fuel the manufacture of carbohydrate (see p.262).

Scavengers eat the soft tissue

**SCAVENGER STAGE**

Worms and crustaceans

Invertebrates feed on the remains

**OPPORTUNISTIC STAGE**

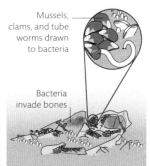

Mussels, clams, and tube worms drawn to bacteria

Bacteria invade bones

**SULFOPHILIC STAGE**

**Pigmented blue**

The mandarinfish (*Synchiropus splendidus*), found in shallow inshore reefs of the Western Pacific, is one of only two known vertebrate species whose blue colour comes from a pigment, rather than structural scattering of light. The striking blue coloration is used as a warning to predators that it is protected by a powerfully unpleasant-smelling toxic mucus secreted through the smooth, scaleless skin.

# blue under water

Vivid colours of fish on a coral reef help to signal species identity in this crowded community. Blue and yellow are the most common: they contrast well and transmit far under water, while red is effective in shallower depths. Much of this colour comes from pigments that absorb some wavelengths of light and reflect others. But some – notably blue – is structural. This happens when objects, such as microscopic crystals in a fish's skin, bend wavelengths to different degrees – in the way raindrops produce a rainbow.

## PIGMENTARY AND STRUCTURAL BLUES

Black, brown, yellow, and red pigments are produced in a fish's skin by cells called chromatophores. So far, blue-pigmented chromatophores (or cyanophores) are known only in the mandarinfish and a close relative. Pigments absorb wavelengths of light, except for the colour that we see, which is reflected back. Other skin cells, called iridophores, contain guanine crystals. As light passes through the crystals, the wavelengths bend to varying degrees. Shorter blue wavelengths get scattered back and intensified, so the skin appears blue. This explains the blue colour of other kinds of coral reef fish.

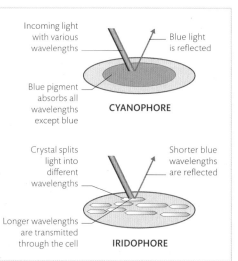

Incoming light with various wavelengths

Blue light is reflected

Blue pigment absorbs all wavelengths except blue

**CYANOPHORE**

Crystal splits light into different wavelengths

Shorter blue wavelengths are reflected

Longer wavelengths are transmitted through the cell

**IRIDOPHORE**

# inflatable bodies

The related puffer fish, balloon fish, and porcupine fish all share the startling defensive strategy of rapid inflation of the body to a near spherical form in order to deter attacks from larger animals. They also produce potent toxins. However, these defences sometimes come too late – large, quick predators may snap up a fish before it has fully inflated, which can result in the death of both the fish and the predator. The unrelated swell sharks adopt a similar strategy, but they don't achieve the extreme roundess seen in the others.

**Difficult mouthful**
The body of the long-spine porcupine fish (*Diodon holocanthus*) is covered in rows of scales that extend into needle-like spines. In a relaxed fish, the spines lie flat against the body, but when the fish swells, the skin tautens and the spines are pulled erect, increasing the deterrent to predators.

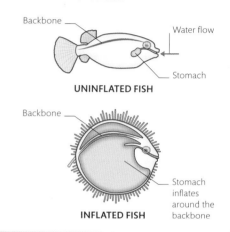

**GULPING WATER**
When frightened, puffer fish and porcupine fish transform by rapidly swallowing water into the stomach, which expands like a balloon, filling all available space in the body cavity. Powerful sphincter muscles at either end of the stomach prevent liquid and gas passing into the gut during inflation. When the perceived threat is passed, the water is regurgitated or belched back out of the mouth and the fish deflates.

Backbone — Water flow

— Stomach

**UNINFLATED FISH**

Backbone

Stomach inflates around the backbone

**INFLATED FISH**

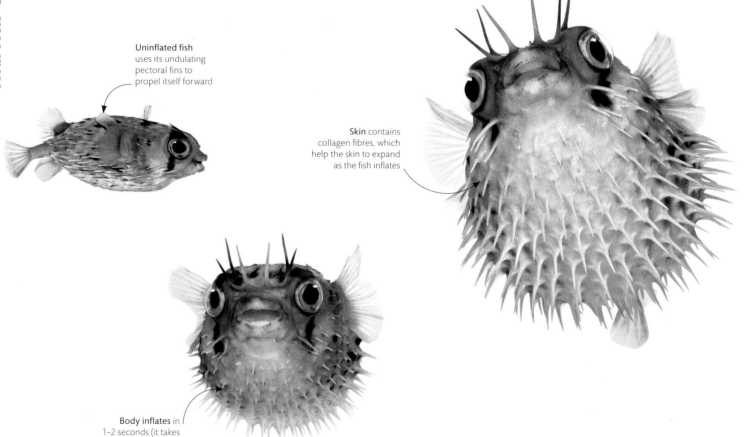

**Uninflated fish** uses its undulating pectoral fins to propel itself forward

**Skin** contains collagen fibres, which help the skin to expand as the fish inflates

**Body inflates** in 1–2 seconds (it takes slightly longer to deflate)

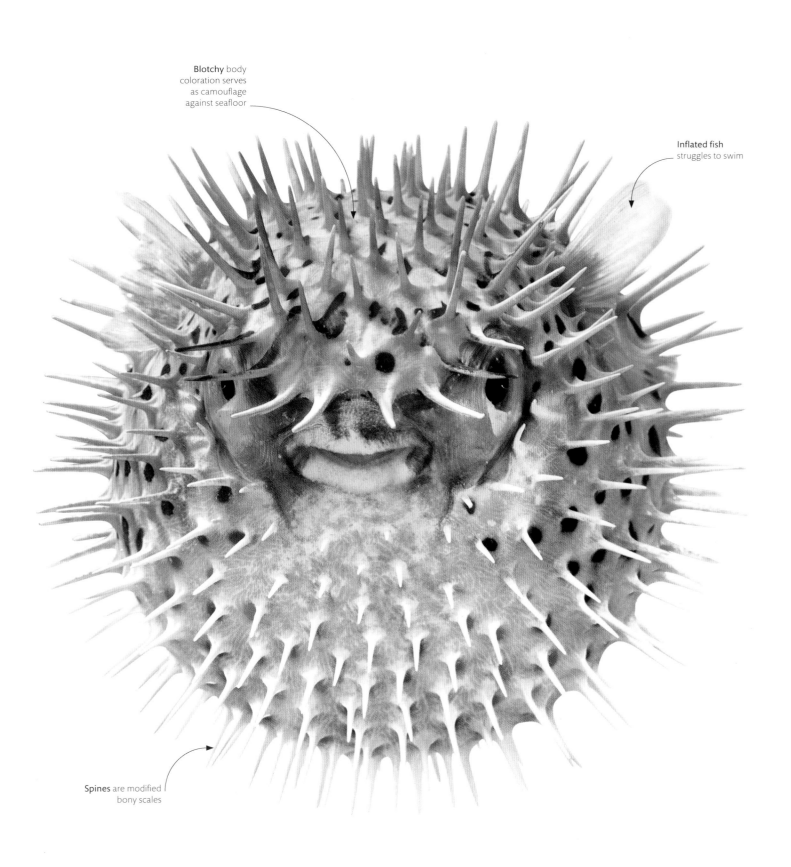

**Blotchy** body coloration serves as camouflage against seafloor

**Inflated fish** struggles to swim

**Spines** are modified bony scales

**Dugong Hunt (1948)**
In this bark painting of fishermen in northern Australia's Gulf of Carpentaria, a dugong – a sea mammal especially valued by Aboriginal people – is pulled from the depths. The work is attributed to the Aboriginal artist Jabarrgwa "Kneepad" Wurrabadalumba.

*the ocean in art*

# aboriginal sea stories

The art of Aboriginal peoples in Arnhem Land, in Australia's Northern Territory, reflects their ancient connection with animals in coastal waters and rivers. Traditionally, Aboriginal peoples use art (and oral literature) rather than the written word to pass on their knowledge and tell stories. Works may convey practical skills, such as fishing, or legends of cultural or spiritual significance, often using symbols passed down through millennia.

**Barramundi (20th century)**
Characteristic of western Arnhem Land art, a pair of barramundi are painted on stripped bark in the traditional X-ray style. The skeleton and organs are mapped out inside the outline of the fish.

**Dreamtime Spirits and Fish (20th century)**
The rock art sites in Kakadu National Park are decorated with paintings and drawings dating back to *c.*26,000 BCE, as well as more recent works. As is typical of Aboriginal rock art, many artists have retold stories by painting over older work. This painting of X-ray-style fish with ceremonial figures is one of 600 works at 46 sites by the artist Najombolmi.

Of all the fish featured in Aboriginal art, barramundi reign supreme. For thousands of years, they were painted on rocks, in sites that are now protected by the Kakadu National Park. Today, the fish is widely reproduced on paper and in bark paintings – an artform that has become widespread in the past 70 years. The bark is stripped from gum trees, then heated and flattened to create a surface for natural pigments.

In western Arnhem Land, subjects are depicted in X-ray style (see left). In the east, sea creatures and fishermen are rendered with intricate cross-hatching, known as *raark*, using reeds or human hair like a brush to create fine lines. In the northeast, the bark paintings of the Yolŋu people tell tales of seafaring ancestors and legendary sea creatures, and impart centuries of knowledge as the caretakers of the coast.

After a barramundi fishing party illegally encroached on a sacred site in 1996, Yolŋu artists in Yirrkala produced a set of paintings known as the Saltwater Barks, to explain their ancient traditions and assert ancestral rights. In 2008, the High Court of Australia confirmed that the area's Aboriginal owners have exclusive access rights to tidal waters along 80 per cent of the Northern Territory coastline.

❝ By painting ... we are telling you a story ... We use our knowledge to paint from the ancient homelands to the bottom of the ocean. ❞

YOLŊU ELDER, *SALTWATER: YIRRKALA PAINTINGS OF THE SEA COUNTRY,* 1999

## Holding fire

The stinging cells (nematocysts) on anemone tentacles are usually triggered by contact with another animal. To avoid being stung, the maroon clownfish (*Premnas biaculeatus*) secretes a body covering of mucus, to which the anemone does not react.

### SECRETING MUCUS

The mucus secreted by various fish organs serves many purposes, not least acting as a barrier to pathogens. Mucus contains a cocktail of carbohydrates and proteins, which varies according to the organ and species. Some evidence suggests that clownfish skin mucus appears to lack the proteins that anemones associate with potential threats or meals, and so contact with the fish does not trigger a stinging response.

**MUCUS CELLS IN SKIN**

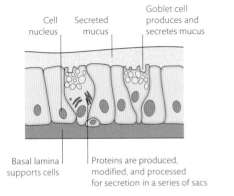

Cell nucleus

Secreted mucus

Goblet cell produces and secretes mucus

Basal lamina supports cells

Proteins are produced, modified, and processed for secretion in a series of sacs

# mutual protection

Several fish species live in mutually beneficial relationships with anemones, most notably clownfish. The fish gain protection from predators for themselves and their eggs among the anemones' stinging tentacles. The anemones benefit because the fish chase away anemone-eaters, drop food, and clear the anemones' tentacles of waste and debris. In addition, nutrients from the clownfish droppings boost the growth of algae (zooxanthellae), which live in even closer association inside the anemone, using light energy to make food.

Larvae hatch a week after the eggs are fertilized by the male

### Safe nursery

Clownfish embryos develop from eggs laid close to anemones, and the young fish are resistant to anemone stings from the moment of hatching. All clownfish begin life as males. The oldest and largest adults switch sex to become females.

# sex changer

Many fish change sex during their lifetime, which means that an individual can breed as both a male and a female at different stages of its life. These fish are called sequential (or non-simultaneous) hermaphrodites. Most of these sex changers, including several species of wrasse, switch from females to males (protogyny). However, some species switch the other way (protandry), including the common clownfish and, according to some experts, the ribbon moray (*Rhinomuraena quaesita*).

Mostly yellow individuals are thought to be females

The wide gape is a threat posture typical of moray eels

**The female stage**
A major advantage of protandry, in which a male increases in size and becomes female, is that the larger females have greater bodily resources to invest in egg production, which requires a great deal of energy.

**Watching and waiting**
All morays spend long periods hidden in crevices with the head protruding. Relatively large eyes, expanded nostrils, and sensitive tentacles help the ribbon moray sense prey, which it seizes with its two sets of jaws – one in the mouth, another in the throat. The ribbon moray is sometimes known as the leaf-nosed moray for its distinctive snout.

Bright blue body coloration of male

Nostrils flare into
delicate, trumpet-
shaped structures

Tip of lower jaw
bears three short,
fleshy tentacles

## SIZE AND COLOUR VARIATIONS

Ribbon morays were once regarded as
several species, because of their striking
colour variation. However, some experts
think that they may develop from small
black juveniles into blue adult males
and then into large yellow females.

Can grow up
to 130 cm
(51 in) long

Typically less
than 65 cm
(26 in) long

Typically
65–95 cm
(26–37 in)
long

JUVENILE

ADULT
MALE

ADULT
FEMALE

### The male stage

As the ribbon moray matures, the jet black of the
juvenile is thought to change to a distinctive, vivid
electric-blue colour. Current thinking suggests this
may be the male stage of its life cycle.

Bright blue coloration,
except on the dorsal
fin, snout, and lower
jaw, which are yellow

Tall dorsal fin
gives the body
a ribbonlike
appearance

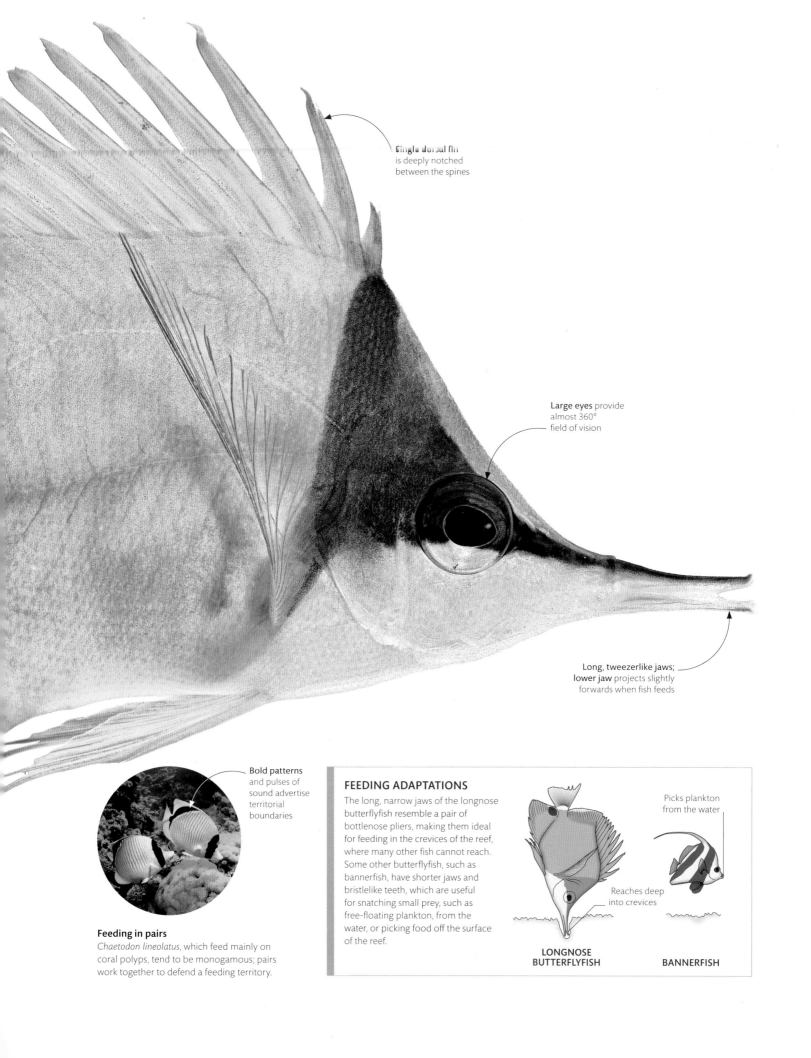

**Single dorsal fin** is deeply notched between the spines

**Large eyes** provide almost 360° field of vision

**Long, tweezerlike jaws; lower jaw** projects slightly forwards when fish feeds

**Bold patterns** and pulses of sound advertise territorial boundaries

**Feeding in pairs**
*Chaetodon lineolatus*, which feed mainly on coral polyps, tend to be monogamous; pairs work together to defend a feeding territory.

## FEEDING ADAPTATIONS

The long, narrow jaws of the longnose butterflyfish resemble a pair of bottlenose pliers, making them ideal for feeding in the crevices of the reef, where many other fish cannot reach. Some other butterflyfish, such as bannerfish, have shorter jaws and bristlelike teeth, which are useful for snatching small prey, such as free-floating plankton, from the water, or picking food off the surface of the reef.

Picks plankton from the water

Reaches deep into crevices

**LONGNOSE BUTTERFLYFISH**

**BANNERFISH**

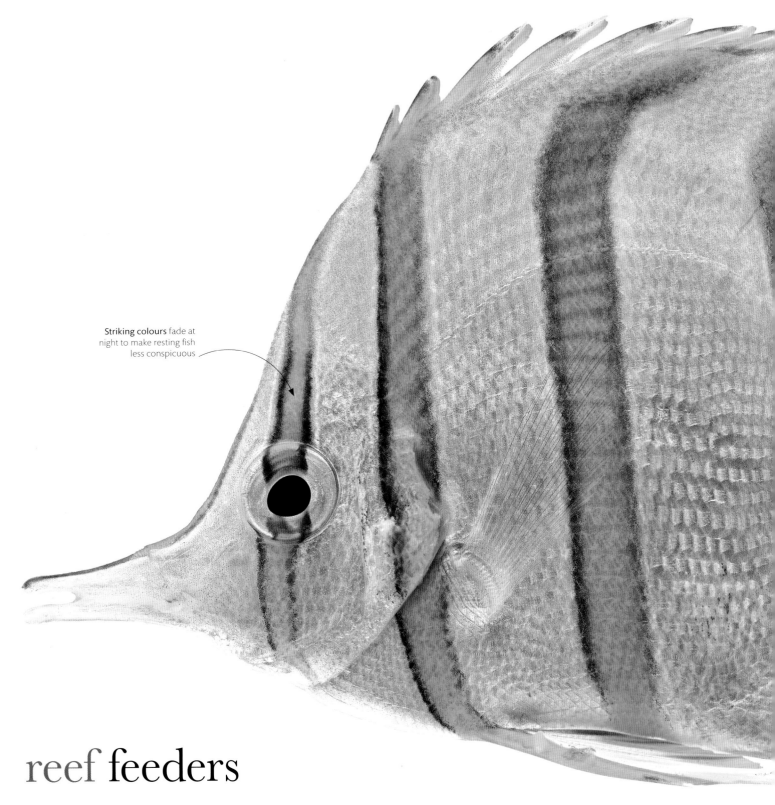

Striking colours fade at
night to make resting fish
less conspicuous

# reef feeders

The 5,000 or more species of fish that live on coral reefs feed in many different ways. This helps them divide up the available resources and avoid competition. Only about 130 eat coral polyps directly, while many more take tiny invertebrates hidden among the coral rock. Dominant among the polyp-eaters and rock-probers are the butterflyfish. Their tiny protractile mouths, armed with comblike teeth, are perfect for foraging among crevices or rubble, from where they extract their miniature prey.

**Dancing colour**
Butterflyfish are named for their colourful appearance and the way they flutter above the reef in "clouds". The yellow longnose butterflyfish (*Forcipiger flavissimus*, opposite) and the smaller copperband butterflyfish (*Chelmon rostratus*, above) are both found in reefs in the Indian and Pacific oceans.

**Spines** of elegant unicornfish are fixed in an open position, while those of most other surgeonfish are retractable

**Dorsal fin** is yellow in *Naso elegans* and mostly black in the otherwise almost identical *N. lituratus*

**Sharp spines** at base of tail project outwards from body

**Body coloration** brightens with age: young specimens are grey; older individuals have a yellow chest

**Sharp end**
Like its close relatives, the elegant unicornfish (*Naso elegans*) possesses at least two forward-pointing spines on either side of the tail. A flick of the tail can inflict serious wounds.

# bladed fish

Most fish are covered in scales, which are small, usually bony plates that grow in the skin. While all scales offer some form of protection, they vary greatly in structure and composition. Some groups have evolved modified scales that serve a particular function. Surgeonfish, including unicornfish and tangs, are distinguished by having large scales on the tail, modified into sharp spines, or scalpel-like blades. These usually lie flat, and the fish raises them when disturbed; however, in some species they are fixed in a raised position.

**Defensive weapons**
Like all surgeonfish, the elegant unicornfish (*Naso elegans*), sometimes called the lipstick tang, uses its spines as a form of defence. It is rarely aggressive towards other fish and is quite confident, perhaps because of its built-in weapons.

**Large eyes**
provide excellent colour vision

**Small mouth**, with orange "lips", contains a single row of sharp teeth used to shred seaweed

## Small prey species

These fish, especially the hidden bottom-dwellers (cryptobenthics), are the some of the smallest and most plentiful and account for almost 60 per cent of fish consumed by larger species. Cryptobenthic larvae remain close to their parents' reefs, increasing their chances of survival, so the fish supplies are constantly replenished.

Translucent cryptobenthic fish that is no more than 2.5 cm (1 in) long

Long, tapered body can be 5 cm (2 in) in length

Skin is smooth and scale-less, but is protected by a layer of mucus

**URCHIN CLINGFISH**
*Diademichthys lineatus*

**GLASS GOBY**
*Coryphopterus hyalinus*

**MAGENTA DOTTYBACK**
*Pictichromis porphyrea*

## Foragers

The water column zone of a reef sees slightly larger fish that feed on the smaller bottom-dwellers, coral polyps, plankton, and micro-invertebrates. They also clean the reef and some, for example, the surgeonfish, also consume algae, which helps to ensure sunlight can reach the coral for photosynthesis.

Male has bright red dorsal fin that it erects during courtship

Laterally compressed shape enables fish to swim through narrow gaps in the reefs

Head has distinctive dark stripe on the snout and electric blue lines at the eyes

**BULLSEYE CARDINALFISH**
*Ostorhinchus fleurieu*

**FLAME ANTHIUS**
*Pseudanthias ignitus*

**COPPERBANDED BUTTERFLYFISH**
*Chelmon rostratus*

## Visiting predators

The largest and most active predators of the reef's food chain are the visitors that feed voraciously on fish, crustaceans, and plankton, as well as benthic organisms such as algae, echinoderms, molluscs, tunicates, sponges, and hydrozoans. They range in size from the larger sharks and rays, to huge schools of smaller species such as triggerfish and snappers.

Venomous dorsal spines are raised to deter would-be predators

Snout is long and pointed with large mouth; upper canine teeth are sometimes visible when mouth is closed

Enlarged scales above the base of pectoral fin, behind the gill, are used to produce sound

**ORANGE-LINED TRIGGERFISH**
*Balistapus undulatus*

**RED LIONFISH**
*Pterois volitans*

**SCHOOLMASTER SNAPPER**
*Lutjanus apodus*

# reef fish

**Lower jaw** carries a pair of venomous canines, used in defence

**STRIPED BLENNY**
*Meiacanthus grammistes*

Coral reefs cover scarcely 0.1 per cent of the world's oceans, but are home to around one-third of all described species of marine fish. The three-dimensional structure of a reef – as complex as a rainforest on land – has provided a rich variety of opportunities for fish to evolve in a multitude of different ways, helping them to coexist and interact with the other life around them.

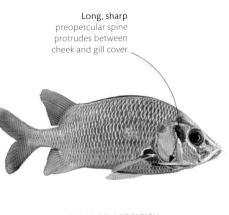

**Long, sharp** preopercular spine protrudes between cheek and gill cover

**SABRE SQUIRRELFISH**
*Sargocentron spiniferum*

**Colour-changing reef dweller**
Streaked with blue over a pale brown ground, the elongate surgeonfish (*Acanthurus mata*) is capable of changing the background colour to dark brown. The species inhabits steep slopes and rocky bottoms and is often seen in large schools. This surgeonfish feeds almost entirely on zooplankton, and can live for more than 20 years.

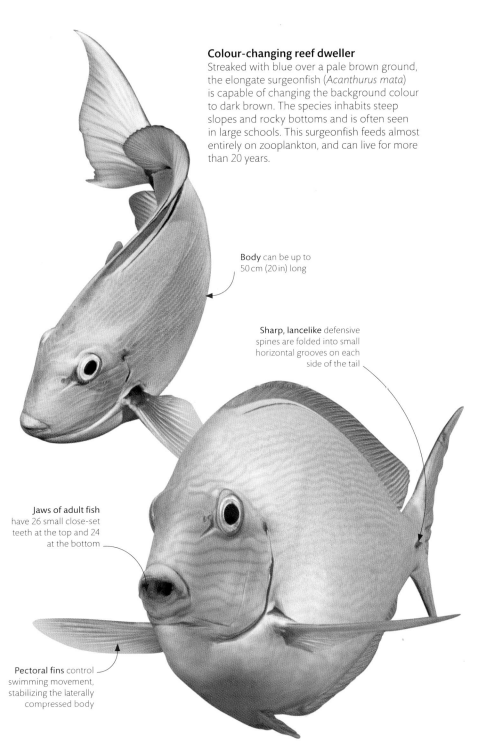

**Body** can be up to 50 cm (20 in) long

**Sharp, lancelike** defensive spines are folded into small horizontal grooves on each side of the tail

**Upper jaw** has outer row of strong canines and an inner band of small teeth

**Jaws of adult fish** have 26 small close-set teeth at the top and 24 at the bottom

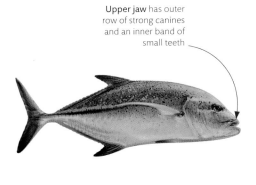

**BLUEFIN TREVALLY**
*Caranx melampygus*

**Pectoral fins** control swimming movement, stabilizing the laterally compressed body

Fleshy tassels (cirri) may
mimic the white polyps that
project from a gorgonian coral

# cryptic fish

Coral reefs and shallow seas are visually complex
environments, and many species take advantage
of the intricate mix of colour and texture to hide from
predators and prey. Ghost pipefish are true masters
of the art of camouflage and crypsis, with different
species specializing in mimicking the appearance
and behaviour of diverse reef-dwelling organisms.

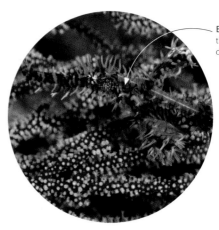

By staying motionless,
the pipefish perfects the
deception

**Perfect match**
The red, elongated body and
spreading fins of a harlequin
ghost pipefish bear a very
close resemblance to the
branching colony of a
gorgonian coral.

## Head down

A head-standing posture may help
a harlequin ghost pipefish blend in
with vertical branches of coral.
Other species of ghost pipefish
adopt a similar behaviour when
hiding in seagrass.

**Eyes** on side of head
rotate in sockets,
providing all-round
vision even while the
fish is motionless

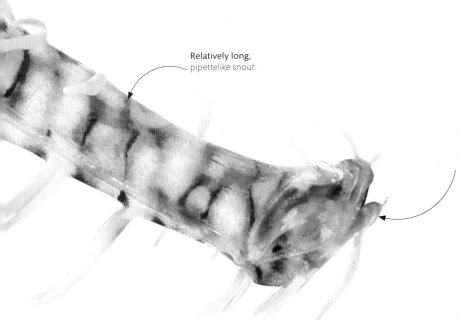

**Relatively long,**
pipettelike snout

**Tiny, toothless mouth**
at the tip of the snout
sucks in prey (typically
tiny crustaceans)

## Colour shift

The harlequin ghost pipefish (*Solenostomus
paradoxus*) changes its appearance to be as
inconspicuous as possible at each stage of
life. The transparent larvae start off in open
water, where they live unobtrusively among
plankton. On maturity, they live on the reef,
where they are surrounded by gorgonian
corals and crinoids and adopt strong colours
and textures and a ragged outline.

# specialized fins

About 99 per cent of all bony fish, of which there are around 30,000 species, are classified as ray-finned fish; this makes them the most species-rich class of vertebrates on Earth. All members of this group have fins made up of fine webs of skin supported by bony rays. The adaptability of this basic structure contributes to the group's success, with size, shape, and function varying to give fish the specialist abilities that set species apart. In triggerfish, the first dorsal fin is small and triangular and contains three thick, spinelike rays. If the fish is relaxed, the spines lie flush with the back, but if it feels threatened, it raises its spines so they are perpendicular to the body.

First dorsal fin, containing three spines, is raised in self-defence

### LOCKING MECHANISM

If a triggerfish feels threatened, it raises the spines in its first dorsal fin. The smaller secondary spine then slots into a groove on the back of the larger front spine, locking it in place. The front spine cannot be lowered until the secondary spine has folded back.

Front spine is locked in upright position

Second spine acts as trigger

Both spines fold down together

**DORSAL FIN SPINES OF A TRIGGERFISH**

### Dual-purpose spine

When the reef-dwelling clown triggerfish (*Balistoides conspicillum*) is targeted as prey by larger species, the lockable spine is an important defence mechanism. In addition to making the triggerfish difficult to swallow, it can be used to wedge the fish in a crevice where the predator cannot reach it.

Second dorsal fin ripples
to provide source
of propulsion

Caudal fin used in short
bursts of fast swimming,
for example when the
fish flees a predator

Anal fins help
to propel fish
through water

Pelvic fins reduced
to a pair of stout
ventral spikes

**Venus in a half-shell (1st century CE)**

In the central peristyle (courtyard) of the House of Venus Marina, in Pompeii, a dazzling fresco panel of Venus takes centre stage. Rising fully formed from her half-shell, she is carried forward by the wind caught in her red mantle, and is attended by sea nymphs and a dolphin.

*the ocean in art*

# seas of antiquity

**An ocean floor (1st century CE)**

A floor mosaic from a Roman caldarium (hot bath) features African fishermen chasing fish and sea monsters. It was recovered from the House of Menander in Pompeii.

**The sea personified (2nd–3rd century CE)**

In Graeco-Roman mythology, Pontus or Oceanus personified the sea and fathered a panoply of sea gods and sea life. In this detail from a Roman mosaic in Utica, Tunisia, the watery god, with a beard of waves, seaweed hair, and lobster-claw horns, is attended by *erotes* (winged cupids) riding on the backs of dolphins.

The sea played a key role in the two ancient civilizations that underpinned much of Western culture. The city-states of ancient Greece fringed the Aegean and extended along the Mediterranean coast, until they came under Roman rule in 146 BCE. At its peak, the Roman empire controlled all the lands around the Mediterranean, and many beyond. For both civilizations, the sea was a major source of inspiration in their mythology and art.

The perfection of classical Greek art survives in statues, sculpted friezes, and mosaics, but rarely in paint. The few frescoes that remain comprise painted tombs in Paestum, southern Italy, dating from the 5th century BCE, and fragments from the much earlier Minoan civilization on the Greek island of Crete. Two in particular honour the sea: in Paestum, a lone diver flexes his body as he swoops down to meet the water; in Crete, dolphins, painted *c.*1500 BCE and restored in the early 20th century, play across the walls at the palace of Knossos.

Further insights into the art of ancient Greece have emerged through Roman art, largely because the expanding empire became a melting pot for the cultures that preceded it. Looted classical treasures, such as bronze statues and paintings of mythological scenes, were prized and preserved through faithful copies. Hellenistic gods were renamed and their stories replicated in the frescoes and mosaics that adorned the homes and gardens of wealthy Romans.

Sea-life mosaics were favoured for bath houses and courtyard pools, especially in the luxury villas of Pompeii, southern Italy. A famous mosaic from the House of the Faun celebrates the coastal bounty of fish, lobsters, eels, and octopuses in a design that was widely used. In coastal colonies in Turkey and North Africa, mosaics featured sea gods and goddesses, such as Poseidon and Amphitrite, and a host of fish-tailed mythical beasts.

Dolphins were popular in mosaics and frescoes as attendants to Aphrodite or playfully circling fishermen and ships. In his *Historia Naturalis* (77–79 CE), Pliny the Elder contributed to their magic with his tale of a dolphin who befriends a boy and carries him to school every day.

> ❝ (The dolphin) does not dread man, as though a stranger to him, but comes to meet ships, leaps and bounds to and fro... ❞
>
> PLINY, *HISTORIA NATURALIS*, 77–79 CE

Jaws are covered
with a mosaic of
hardwearing teeth,
which are continually
replaced throughout life

## TEETH AND DIET

The jaws and teeth of fish show
a huge variety of specialist forms
and functions. For example, the
teeth of the queen parrotfish
(*Scarus vetula*), which is a herbivore,
are completely blunt, with a highly
resilient, replaceable surface that
is used to scrape the surface of
rock and coral without damaging
the underlying bone. In contrast,
the carnivorous northern red snapper
(*Lutjanus campechanus*), which feeds
on small fish and crustaceans, has
needle-sharp teeth that are used
as weapons for stabbing and
snagging its prey.

Blunt, fused
teeth

**HERBIVORE (QUEEN PARROTFISH)**

Sharp,
pointed teeth

**CARNIVORE
(NORTHERN RED SNAPPER)**

**Excreting sand**

As the steephead parrotfish (*Chlorurus microrhinos*) feeds, it gouges particles of reef and rock from the substrate, which it ingests at the same time as its food. The particles pass through the gut undigested and are then excreted as fine white sand.

**Robust beak** and powerful jaws are typical of excavating parrotfish

# reef scrapers

The sunlit surfaces of rocky and coral reefs make an ideal growing substrate for algae, which in turn provide an abundant food resource for other reef-dwellers such as parrotfish. These colourful fish, which are members of the wrasse family, have teeth and jaws that have evolved into a beaklike structure. Some species use their teeth to scrape algae from the reef surface, and in others the beak is so tough it can gouge into the reef. These so-called excavating parrotfish supplement their algal diet with animal matter in the form of coral polyps.

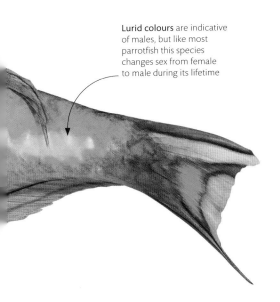

**Lurid colours** are indicative of males, but like most parrotfish this species changes sex from female to male during its lifetime

**Reef resources**

The ember parrotfish (*Scarus rubroviolaceus*, far left) and yellow-finned parrotfish (*S. flavipectoralis*, left) are closely associated with tropical reefs. They feed by scraping algae from the surface of rocks and corals using their beak. As well as providing a vast surface area for feeding, the complex reef structure also provides shelter for the fish.

# coastal
# seas

**Rich in nutrients** from ocean upwellings
and land run-off, coastal seas have abundant
plant and animal life. The shallow, light-
drenched habitat is ideal for photosynthesis,
especially for small plants and macroalgae
rooted to the seabed.

# absorbing light

Like all algae, seaweeds make food by photosynthesis, using sunlight's energy to convert carbon dioxide and water into sugar and other nutrients. All seaweeds use pigments to absorb this energy, while their fronds maximize the surface area for light capture. Different pigments absorb different colour wavelengths present in white sunlight. Green chlorophyll predominates, but some seaweeds have accessory pigments that make them brown or red.

**The green colour** of the frond is due to the photosynthetic pigment chlorophyll, which is packed inside microscopic structures called chloroplasts

**The frond's cylindrical midrib** has chloroplasts that carry out photosynthesis; it transports the nutrients produced by photosynthesis to other parts of the seaweed

### Selecting the spectrum
Green seaweeds, such as sea grapes (*Caulerpa lentillifera*), lack the accessory pigments of other coloured seaweeds. Their colour is due largely to chlorophyll, the same pigment used by land plants. Chlorophyll absorbs red and blue wavelengths of light – for use in photosynthesis – but reflects green wavelengths.

**Grapelike ramulus** stores sugars and other nutrients that are generated by photosynthesis; each frond consists of a branchlike midrib carrying rows of ramuli

**Fronds** are stiffened with white mineral

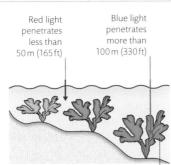

### Red seaweed
In addition to chlorophyll, red seaweeds – such as this *Corallina* – have accessory pigments known as phycobiliproteins, which reflect more red light and absorb more green.

### LIGHT AND HABITAT
Photosynthetic pigments help seaweeds to thrive in different habitats. Blue and green light penetrate water more deeply than other wavelengths, enabling red seaweeds that absorb these colours to grow better at greater depths. But the quantity of pigment, as well as the type, is important too: some seaweeds of other colours have higher pigment levels to absorb more light in deeper water.

Red light penetrates less than 50 m (165 ft)

Blue light penetrates more than 100 m (330 ft)

Deeper seaweeds have more pigment for absorbing available wavelengths

**SEAWEED DEPTH DISTRIBUTION**

# giant kelp

Forming underwater "forests" rising more than 30 m (100 ft) from the seabed, giant kelp (*Macrocystis pyrifera*) is a type of algae that acquires food by photosynthesis. It is found off the North American Pacific coast from Canada to California, and in temperate regions of all southern-hemisphere oceans.

Giant kelp is the largest living seaweed – the term for large algae growing on ocean coastlines. They are supported by water, which enable some seaweeds to grow to enormous lengths. Giant kelp is thought to have the fastest linear growth rate on Earth, and can increase its length by up to 61 cm (2 ft) per day. In clear, nutrient-rich waters where the sea temperature is 5–20° C (42–68° F), it can survive at depths of 53 m (175 ft), but it reproduces only if the water temperature is below 18–20° C (65–68° F).

### A forest for carbon capture
Giant kelp forests, like this one off the coast of California where blacksmith damselfish shelter, absorb vast amounts of carbon dioxide. They act as natural carbon sinks in a similar way to terrestrial forests and lower seawater acidity levels.

Giant kelp anchors itself to the seafloor using holdfasts – rootlike attachment organs that form a tangled mass around rocks, rubble, and debris on the seabed (see p.27) – to prevent it drifting ashore. Rocky coastlines make ideal fastening places. A primary stipe, or main stem, rises towards the surface, sending out branching side stalks with leaflike blades, each with a pneumatocyst, or gas bladder, at its base that keeps it afloat. At the surface, the kelp covers the water in a type of canopy. Unless compromised by warming seas or overgrazed by sea urchins, kelp can live up to seven years.

Kelp forests provide a key habitat for many animals. Holdfasts shelter invertebrates, fish and their larvae, and shelter birds and sea otters during storms. Sharks, sea lions, and seals are also attracted to these rich hunting grounds.

**Pneumatocysts** hold the blades upright, enabling kelp to take in sunlight for photosynthesis

### Staying afloat
Balloonlike pneumatocysts are filled with a mixture of carbon dioxide, oxygen, and nitrogen that giant kelp cells both produce and absorb from surrounding sea water.

# underwater meadows

Most true plants cannot grow in the ocean. While the sunlit shallows can be thick with seaweeds and other algae, these are solar-powered organisms that lack true leaves, roots, and flowers. However, seagrasses, such as turtle seagrass (*Thalassia testudinum*), are flowering plants. Rather than merely tolerating the salty ocean environment, they thrive in it: they bloom and set seed below the surface and cover vast areas of the seabed. At a time of global warming, these photosynthesizing green carpets are globally important as efficient stores of atmospheric carbon.

**Asexual reproduction**
Turtle seagrass produces horizontal rhizomes that sprout plant clones, helping the species to form extensive swards – much like grasses do on land.

## SEXUAL REPRODUCTION

In turtle seagrass, genetic diversity comes from cross-fertilization during sexual reproduction. The sexes are separate: each plant produces either male or female flowers. Male flowers release pollen at night in packages of carbohydrate-rich slime. The slime attracts hungry invertebrates, which help disperse the pollen to female flowers. Fertilized flowers form seeds encased in small fruits. The fruits float before settling on the seabed, where each seed completes its germination.

Sticky pollen bundles

Up to five flowers per plant

**MALE**

Single flower per plant

**FEMALE**

Roots anchor seedling to seabed

**SEEDLING**

Seeds sprout while fruit is floating

**FRUIT**

### Tropical seagrass

Turtle seagrass – the favourite food of the green turtle (*Chelonia mydas*) – is a common seagrass of the Caribbean. Neighbouring plants are interconnected clones, but they also reproduce sexually: their seed-packed fruits disperse on the waves and establish new colonies elsewhere.

# producing light

Many soft-bodied, jellylike animals of the open ocean can produce light: a phenomenon known as bioluminescence. Species as diverse as jellyfish, hydrozoans, and comb jellies turn luminous if the water around them is agitated – perhaps to scare predators or attract tiny planktonic prey. The light, produced by a chemical reaction in organs called photophores, is most striking when emitted in dark, deep water or at night nearer the surface.

**Colourless jelly**
Like many vulnerable jellyfish, the thumb-sized crystal jelly (*Aequorea victoria*), a hydrozoan from the surface waters of the eastern Pacific Ocean, has a transparent body. This may help to conceal it from predators against the open water. Its light-producing photophores, which are carried around the rim of its swimming bell, produce a momentary green glow.

**Rim of the bell** contains photophores, but any light produced is only visible in dark conditions

**Transparent bell** pulsates to propel the crystal jelly through the water

**More than 100 tentacles** hanging from the rim of the bell carry nematocysts (stinging cells); they immobilize and catch tiny planktonic prey

### HOW PHOTOPHORES WORK

Bioluminescence typically depends upon a light-producing chemical and an enzyme. When stimulated by a trigger – such as calcium, in the case of the crystal jelly – the enzyme makes the light-producer undergo a chemical change, which releases light. The distinctive green glow of the crystal jelly happens because an additional component – a fluorescent protein – subsequently changes the wavelength of the emitted light from blue to green.

**ARRANGEMENT AND FUNCTION OF CRYSTAL JELLY PHOTOPHORES**

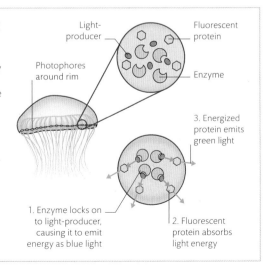

Light-producer

Photophores around rim

Fluorescent protein

Enzyme

1. Enzyme locks on to light-producer, causing it to emit energy as blue light

2. Fluorescent protein absorbs light energy

3. Energized protein emits green light

**Densely packed photophores** glow green and illuminate the rim of the bell

**Glowing green**
Once activated, a cocktail of chemicals in the photophores of a crystal jelly emits green light; each glow lasts for just a few seconds.

**Tubulariae. — Röhrenpolypen.**

# alternating generations

Animals that are attached to the substrate must get their nourishment close to the seabed. Hydrozoans – relatives of anemones, corals, and jellyfish – have polyps with tentacles that may rise only a short distance from where they are growing. But many also produce swimming, jellyfishlike bells, called medusae, as part of their life cycle. Medusae travel far in the water column, where they catch plankton beyond the reach of polyps. By alternating between polyp and medusa generations, food sources are more fully exploited.

**Two generations**
Ernst Haeckel's *Kunstformen der Natur* (1904) depicts more than a dozen species of hydrozoan – some as anemonelike polyps (bottom) and others as medusae (top). Not all have a life cycle that switches between the two stages: some do not become medusae and, like many other ocean invertebrates, experience open water only as microscopic larvae.

**HYDROZOAN SEXUAL LIFE CYCLE**
The hydrozoan *Obelia* grows into an upright branching colony of polyps, some of which generate pods that produce medusae. The free-swimming medusae release sperm and eggs into the water. Larvae develop from the fertilized eggs and eventually settle on the seabed, where they form new polyp colonies. The life cycles of other hydrozoans and related animals vary. Some hydrozoans – as well as corals and anemones – do not have a medusa stage, while certain jellyfish species exist only as medusae.

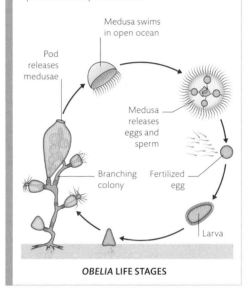

Medusa swims in open ocean

Pod releases medusa

Medusa releases eggs and sperm

Branching colony

Fertilized egg

Larva

*OBELIA* LIFE STAGES

**Polyps in close-up**
The polyps of oaten pipes (*Tubularia indivisa*) – a European hydrozoan that lacks a medusa stage – carry rings of drooping tentacles on stems that extend up to 15 cm (6 in) from their base.

**Tentacles** trap microscopic plankton and transfer them to the mouth at the centre

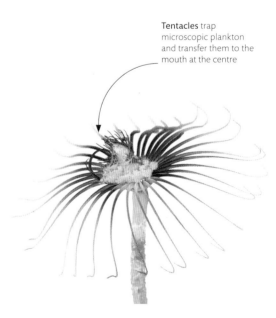

**Gonozooids** are reproductive capsules that release larvae; the larvae develop directly into new polyp colonies

**The Miracles of Mazu (18th century)**
An album of seven ink drawings, now housed in
the Rijksmuseum, Amsterdam, Holland, describes
the sea journey of a Song dynasty deputation to
the Goryeo court in Korea in 1123 CE. The Chinese
ships are under the protection of the sea goddess,
Mazu. She is shown here in her traditional red
robes, hovering on a cloud and carrying an
ivory *Hu* tablet as a symbol of power.

**The Story of the Great Flood (*c.*1870)**
In this depiction of a Hindu legend, Matsya (one of the 10 avatars of the Hindu God Vishnu) appears as half-human, half-fish. Here, he saves Manu (the first man) and the seven sages from the flood.

*the ocean in art*

# asia's sea deities

Thousands of years of art and storytelling reflect humans' longstanding awe of the ocean. In most ancient cultures, the sea's unfathomable depths and unpredictability gave rise to numerous myths and beliefs, in which natural phenomena – such as storms, currents, and fair winds – were attributed to the whims and protective powers of sea gods and goddesses.

In ancient Hindu texts (Vedas), Varuna was the god of the sky, but in later tales he came to symbolize oceans, clouds, and water. In Mysore painting, a classical art form of southern India featuring deities and mythological scenes, Varuna is often pictured riding the *makara* – a hybrid sea creature resembling a crocodile. Varuna also appears in miniatures that illustrate the 5th-century BCE Indian epic *Ramayana*, rising from the ocean to pacify Rama, an avatar of the god Vishnu. According to the legend, Rama's wife, Sita, was held in captivity across the ocean, in Lanka.

In China, from the 11th century on, the success of perilous sea journeys was often attributed to the benificence of the goddess Mazu. Her story varies with the telling, but she was born Lin Moniang in 960 CE in a Chinese coastal town, where she was renowned for her swimming skills. She gained mystical status after she used trancelike powers to save male members of her family from drowning at sea. In 1281, she was given the official title of Tianfei ("Princess of Heaven").

In the 1400s, Chinese admiral Zheng He had his great deeds carved in tablets of granite in two cities in China. He attributed his seafaring success to Mazu/Tianfei, describing her as a "divine light" above the mast, calming the crew in rough seas. In reality, this light was probably St Elmo's fire, a ball of light formed from electricity, which appears on ships at sea in storms. The same goddess is evoked in an 18th-century album of ink drawings (see left), where she is pictured floating above Chinese ships, providing protection.

Mazu worship spread to sailors and fishermen beyond China, and lacquered wooden statues, woodcut illustrations, and painted panels of the goddess are found in temples dedicated to her across the world.

> " When we met danger, once we invoked the divine name, her answer to our prayer was like an echo. "
>
> *ADMIRAL ZHENG HE'S DEDICATION TO MAZU/TIANFEI, LIUJIAGANG TEMPLE,* 1431

Inner whorl
of shorter labial
tentacles transfers
prey to the
central mouth

Tube is made from
specialized stinging cells
and sediment, all bound
together by mucus

## Tube sanctuary

Tube anemones cannot retract their tentacles, but when disturbed they pull themselves into their tube by contracting muscles along the length of their body.

**The tentacles disappear** as the anemone withdraws its body deep into the tube

**Outer whorl of longer marginal tentacles** has stingers for catching prey and self-defence

# living in a tube

Most anemones must make a living attached to a firm surface and can reach no higher than the stretch of their tentacles, but one group has evolved the means not only to live on soft sediment, but even to move up and down on the seabed. Tube anemones live inside vertical, partially buried, feltlike tubes. Instead of being anchored to the bottom, they can slip upwards through the tube or sink deeper inside when danger threatens.

## Emerging tentacles

A fireworks tube anemone (*Pachycerianthus multiplicatus*), from European coastal waters, erupts from the confines of its tube to sweep the water with its tentacles for planktonic prey. A mature specimen can have tentacles that span 30 cm (12 in).

### ANEMONE ANATOMY

All anemones have just one opening to their gut for ingesting food and eliminating waste. The gut is a simple cavity lined with digestive cells. Prey caught and paralysed by the tentacles are transferred to the gut opening, before being digested by enzymes released from the gut lining. The basal disc is attached to a surface in most anemones; in tube anemones, it is free to move up and down.

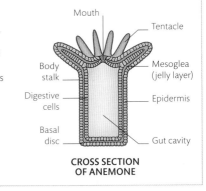

Mouth
Tentacle
Body stalk
Mesoglea (jelly layer)
Digestive cells
Epidermis
Basal disc
Gut cavity

**CROSS SECTION OF ANEMONE**

**Touch-sensitive antenna**, one of three, aids navigation when the fireworm crawls over a reef

**Fleshy projections** called palps, one on either side of the antennae, are used to manipulate food

**Third, central antenna** is shown here curving downwards

**Caruncle** carries chemical sensors and grows as an extension from the first body segment

**Fleshy appendage**, or parapodium, carries sensory filaments called cirri and bundles of bristles known as chaetae; each body segment has a pair of parapodia

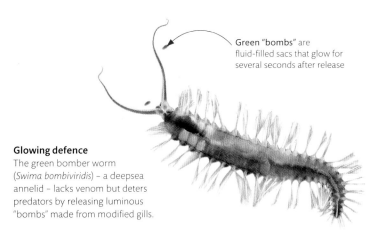

Green "bombs" are fluid-filled sacs that glow for several seconds after release

**Glowing defence**
The green bomber worm (*Swima bombiviridis*) – a deepsea annelid – lacks venom but deters predators by releasing luminous "bombs" made from modified gills.

# stinging bristles

Annelids are segmented worms that include many familiar animals, including intertidal lugworms and land-living earthworms. Their bodies have tiny bristles called chaetae that help with traction when crawling or burrowing. But some marine annelids use their chaetae for defence. Aptly named fireworms have brittle, venom-packed chaetae that break off to deliver a painful sting when the worm is threatened by a predator.

Each tentaclelike sensory **filament**, or cirrus, may be touch-sensitive

Bundle of bristly chaetae – each chaeta is reinforced with tough chitin and impregnated with chalky minerals to make it brittle, so its tip can break

**Stinging worm**
The predatory golden fireworm (*Chloeia flava*) is found along Indo-Pacific coastlines. This annelid tracks victims, such as anemones and sponges, using a fleshy organ called a caruncle that "tastes" the water for chemical traces of prey animals. Venomous bristles protect the animal from shorebirds and worm-eating fish.

---

**AGGRESSIVE VENOM**

Venoms are toxic chemicals that are injected into a victim. Most venomous ocean animals, including fireworms, use their venom defensively, but some deploy it aggressively to catch prey. Cone snails, despite being slow crawlers, successfully hunt fish using a rapid-fire venomous dart, which causes quick paralysis. The dart is a modified radula – a mouthpart used for grazing algae in more conventional snails (see pp.34–35). It is located at the tip of the proboscis, which the snail fires from its mouth.

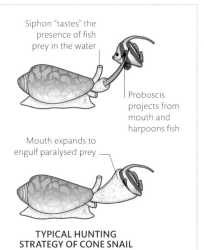

Siphon "tastes" the presence of fish prey in the water

Proboscis projects from mouth and harpoons fish

Mouth expands to engulf paralysed prey

**TYPICAL HUNTING STRATEGY OF CONE SNAIL**

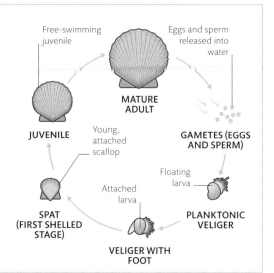

## SCALLOP LIFE CYCLE

It is only in the juvenile and adult life cycle stages that scallops can swim independently. These molluscs are hermaphrodites. Each individual releases both sperm and eggs into the surrounding water. Fertilized eggs develop into planktonic larvae called veligers – a life stage also seen in other oceanic bivalves and snails. After growing a crawling foot, each veliger settles on seagrass or weed, to which it attaches by sticky threads. Here the scallop grows until it is big enough to escape the attentions of crabs and other bottom-dwelling predators. It then develops into a free-living, sexually mature adult.

Free-swimming juvenile

Eggs and sperm released into water

**MATURE ADULT**

**JUVENILE**

Young, attached scallop

**GAMETES (EGGS AND SPERM)**

Floating larva

Attached larva

**SPAT (FIRST SHELLED STAGE)**

**PLANKTONIC VELIGER**

**VELIGER WITH FOOT**

### Multiple eyes

Unlike other bivalves, scallops have two rows of dozens of eyes that alert them to movement so they can swim away from danger. Each eye has a crystalline lens and a retina for image-processing.

**A scallop eye** contains a reflective, mirrorlike layer, which improves its light-collecting ability

### Escaping danger

Scallops, such as this bay scallop (*Argopecten irradians*) of North America's Atlantic coast, can use their flapping shell valves in an impressive escape response. Their complex eyes, which give sharp resolution for both peripheral and central vision, may help them to recognize approaching predators.

# flapping to swim

As their name suggests, bivalve molluscs have a two-part shell. Hinged across the middle, the shell is closed by muscles that span the hinge and contract to pull the halves, called valves, together. Bivalves relax these muscles to open their shells in order to strain particles of food from the water but are otherwise mostly sedentary. Scallops are exceptional: they can swim in open water by rapidly flapping their valves open and shut.

**Brown and white mottling**
helps to disguise the
cuttlefish against the
sediment of the seabed

**Patches of yellow** are
flashed to startle predators
when this colour-changing
cuttlefish is disturbed

**Mantle** – a sleeve of skin that
envelops the cuttlefish – encloses a
small cuttlebone in the front half of
the body; negative buoyancy helps
to keep the animal near the seabed

**Knoblike papilla** (one of six
paired projections) on the
upper body break up the
outline of the cuttlefish to
help with camouflage

**Buoyancy in miniature**
The tiny, bottom-living flamboyant
cuttlefish (*Metasepia pfefferi*), of tropical
Australia, has a proportionately smaller
cuttlebone than mid-water cuttlefish, so
it regulates its buoyancy across a narrower
range of depths. Its toxic flesh – signalled
by warning colours – helps to offset its
vulnerability to more agile predators.

**The cuttlebone** of an elegant cuttlefish (*Sepia elegans*) extends the length of its body

**Each oblong chamber** is less than 1 mm long

**Cuttlebone**
When cuttlefish die, their buoyant, gas-filled cuttlebones may wash ashore on beaches. A magnified cross section reveals the stacks of hollow chambers inside.

**INTERNAL STRUCTURE**

**Each arm** is equipped with four rows of suckers that handle fish and crustacean prey

**Pink arms** can extend downwards to grip the substrate and be used for "walking" on the seabed

# changing buoyancy

Unlike most fish, which stay buoyant in the water by means of an adjustable gas-filled swim bladder (see p.231), some animals change their buoyancy to sink or float. Cuttlefish do this with an internal shell – their cuttlebone. The cuttlebone's hollow chambers buoy the animal upwards, which enables it to swim in open water. To descend to the seabed, the cuttlefish floods the chambers with liquid, making the animal negatively buoyant and weighing it down so that it can burrow or "walk" on the bottom.

**CEPHALOPOD SHELLS**

Cephalopods include a range of swimming molluscs: nautiluses, octopuses, ram's horns, cuttlefish, and squid. The nautilus has an external shell, which is coiled like that of a snail yet chambered like a cuttlebone for buoyancy control. The ram's horn - a deepwater relative of cuttlefish - uses its internal coiled shell in a similar way. In squid, an internal shell is reduced to a simple, penlike structure, or gladius, for support. Octopuses lack a shell altogether, but they are more flexible as a result.

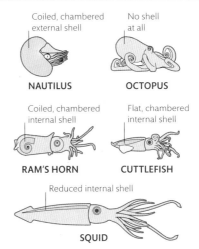

Coiled, chambered external shell

No shell at all

**NAUTILUS**

**OCTOPUS**

Coiled, chambered internal shell

Flat, chambered internal shell

**RAM'S HORN**

**CUTTLEFISH**

Reduced internal shell

**SQUID**

**View of the vortex**
Taken from the International
Space Station in September 2018,
this photograph shows the spiralling
movement of the vortex around the
eye of typhoon Trami in the Pacific.

# hurricanes and typhoons

Violent tropical storms known as hurricanes (in the Atlantic) and typhoons (in the Pacific) are characterized by spiralling winds of up to 350 kph (220 mph), often accompanied by heavy rainfall. They occur when the temperature in the upper 45 m (150 ft) of the ocean exceeds 27°C (80°F). Water temperature reaches this level regularly in the equatorial ocean regions, especially in the Pacific, western Atlantic, and Indian oceans in late summer and autumn. Global warming is increasing the duration, frequency, and intensity of such storms. Hurricanes and typhoons can cause serious damage in coastal areas but abate as they move over land and lose contact with the forces that drive them.

## INSIDE A HURRICANE

An increase in surface temperature causes vast amounts of water vapour to evaporate from the sea surface, which creates huge convection towers up to 15 km (10 miles) high. Low pressure at the surface pulls in more air, creating cloud systems and spiralling thunderstorms. Coriolis force (an effect of Earth's rotation) causes the winds to spin, and as the storm grows it becomes self-sustaining.

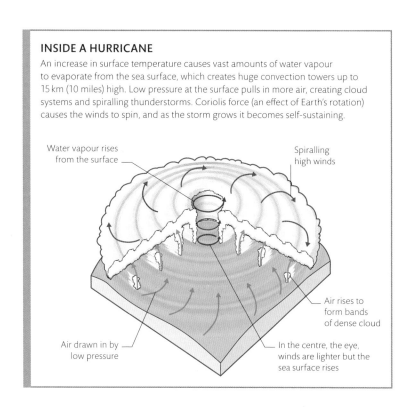

Water vapour rises from the surface

Spiralling high winds

Air rises to form bands of dense cloud

Air drawn in by low pressure

In the centre, the eye, winds are lighter but the sea surface rises

The fins are pulled **against the body** during jet propulsion, to give the body a more hydrodynamic shape

**Each non-retractable** arm has suckers along its length for gripping prey

**Each retractable** tentacle carries suckers on its spatula-shaped tip

# jet propulsion

Molluscs such as snails and slugs creep slowly on a muscular foot, but some molluscs are able to swim in mid-water. Cephalopods, including squid and octopuses, are the fastest of all mollusc swimmers. They adopt torpedo-shapes to cut through water and can generate significant thrust, notably by shooting a jet of water from a siphon, which propels them with sudden bursts of speed.

**A rippling fin** runs along each side of the squid's mantle

**Fin power**
The fins undulate to push against the water. They create most of the thrust for slower, routine swimming, when jet propulsion is not required.

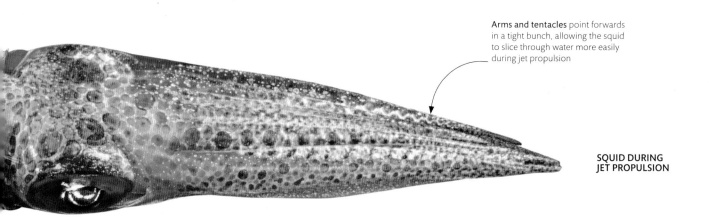

**Arms and tentacles** point forwards in a tight bunch, allowing the squid to slice through water more easily during jet propulsion

**SQUID DURING JET PROPULSION**

**Extended fins** along the side of the body give better control of movements during slow swimming

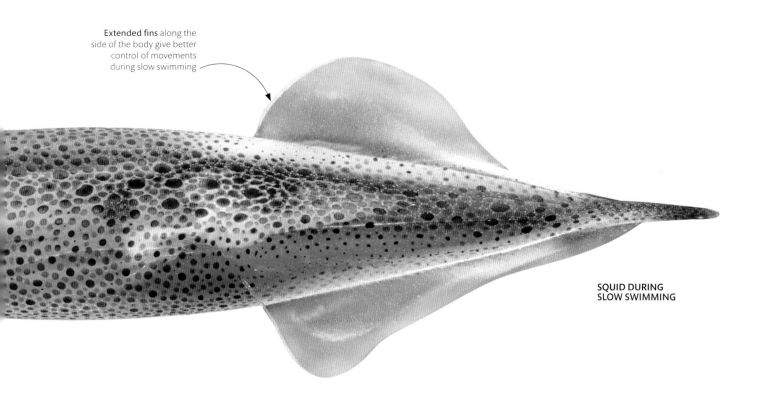

**SQUID DURING SLOW SWIMMING**

## CREATING THE JET

All molluscs have a skinlike mantle – a structure that produces a shell in snails but which envelopes the body like a sleeve in cephalopods. Muscles in the mantle wall dilate the underlying cavity, which swells with water, then constrict to eject a water jet out of a movable siphon. By directing the siphon – and the jet's thrust – forwards or backwards, the squid can control the direction in which it moves.

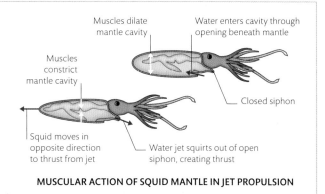

Muscles dilate mantle cavity

Water enters cavity through opening beneath mantle

Muscles constrict mantle cavity

Closed siphon

Squid moves in opposite direction to thrust from jet

Water jet squirts out of open siphon, creating thrust

**MUSCULAR ACTION OF SQUID MANTLE IN JET PROPULSION**

## Speedy predator

Jet propulsion helps the European squid (*Loligo vulgaris*) to cover long distances quickly. It is reserved for escaping predators and catching fast-moving prey: the eight arms and two retractable tentacles are used to grab fish and crustaceans. But jet propulsion does not allow the fine control needed for turning and other manoeuvres, when the squid relies on using its fins instead.

# nudibranchs

These colourful molluscs are found in shallow and deep marine waters the world over. There are more than 3,000 known species; some crawl on the seabed, while others live in the water column. Nudibranchs are divided into two groups – aeolids and dorids. Both types are hermaphrodites, with individuals having male and female sexual organs, but they cannot self-fertilize.

**Collage of colour**
This transluscent white aeolid *Coryphella verrucosa* is found in the tidal streams and sheltered parts of the northern Atlantic and Pacific oceans. Eggs are spawned in a distinctive coiled white string, and like all nudibranchs, it loses its shell at the larval stage. This species can grow to a length of 3.5 cm (1 1/3 in).

## Aeolid nudibranchs

This group have sense organs (rhinophores) at their heads and breathe using tentaclelike outgrowths called cerata that cover their backs. In some, their bright colours act as camouflage; for others, they warn predators of their toxicity. Several species feed on larger, well-armed prey like jellyfish, ingesting their unfired nematocysts, and storing them in special pouches at the tips of the cerata - so the cerata play a role in defence as well as gaseous exchange.

Rhinophores, or sense organs, have distinctive white tips

Surfaces of side (oral) tentacles and rhinophores are covered with wartlike bumps

Red oesophagus, or gullet, visible under translucent skin

**EDMUNDSELLA PEDATA**
2 cm (¾ in) long

**BERGHIA NORVEGICA**
3 cm (1 ⅛ in) long

**FACELINA BOSTONIENSIS**
5.5 cm (2 ⅛ in) long

## Dorid nudibranchs

This group are mostly smooth bodied. They have two rhinophores, or sense organs, at the head and tufts of feather-like gills towards the back of the mantle, which they use for breathing. In some species, these gills can be retracted into a special pouch. Like aeolids, they are grazing carnivores and feed on a variety of animals, including other nudibranchs.

Orange-tipped body processes can be up to 2 cm (¾ in) long

Mantle is covered with small, rounded projections

Feathery, branched gills surround the anus

**LIMACIA CLAVIGERA**
2 cm (¾ in) long

**GONIODORIS CASTANEA**
4 cm (1 ½ in) long

**NEMBROTHA KUBARYANA**
12 cm (5 in) long

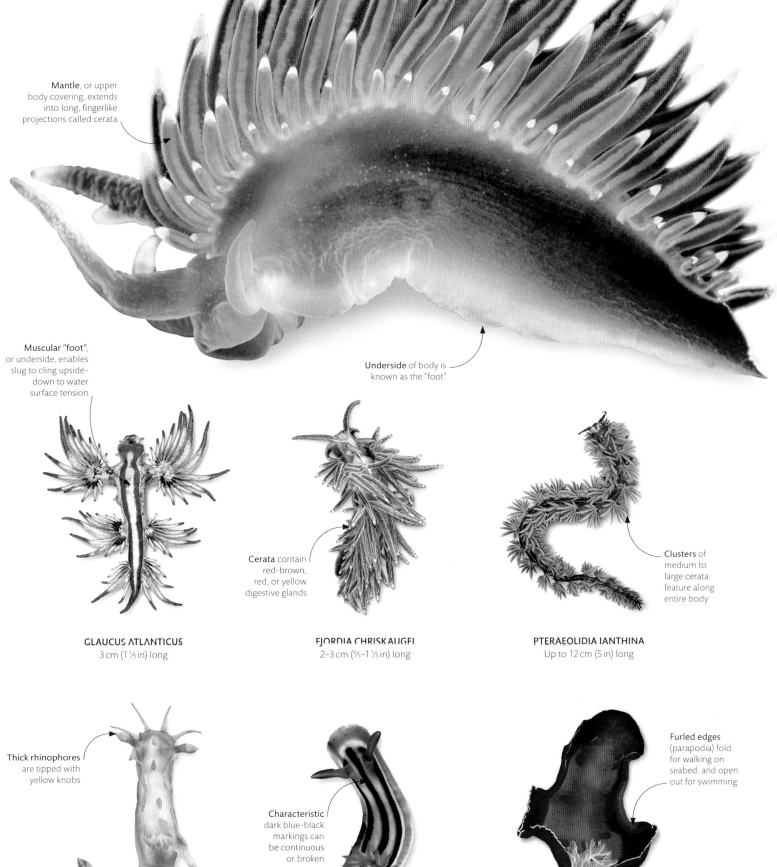

**Mantle**, or upper body covering, extends into long, fingerlike projections called cerata

**Muscular "foot"**, or underside, enables slug to cling upside-down to water surface tension

**Underside** of body is known as the "foot"

**Cerata** contain red-brown, red, or yellow digestive glands

**Clusters** of medium to large cerata feature along entire body

**GLAUCUS ATLANTICUS**
3 cm (1 ⅕ in) long

**FJORDIA CHRISKAUGEI**
2–3 cm (⅘–1 ⅕ in) long

**PTERAEOLIDIA IANTHINA**
Up to 12 cm (5 in) long

**Thick rhinophores** are tipped with yellow knobs

**Characteristic** dark blue-black markings can be continuous or broken

**Furled edges** (parapodia) fold for walking on seabed, and open out for swimming

**POLYCERA QUADRILINEATA**
4 cm (1 ½ in) long

**CHROMODORIS ELISABETHINA**
5 cm (2 in) long

**HEXABRANCHUS SANGUINEUS**
40 cm (16 in) long

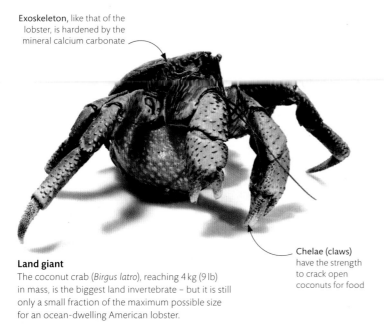

**Exoskeleton**, like that of the lobster, is hardened by the mineral calcium carbonate

**Land giant**
The coconut crab (*Birgus latro*), reaching 4 kg (9 lb) in mass, is the biggest land invertebrate – but it is still only a small fraction of the maximum possible size for an ocean-dwelling American lobster.

**Chelae (claws)** have the strength to crack open coconuts for food

# heavy bodies

For animals, growing big has pros and cons. While a large animal might overcome predators or competitors with brute strength, carrying weight puts strain on the body. Arthropods – jointed-legged animals such as crustaceans – wear their skeleton externally as a suit of armour. A bigger arthropod needs thicker armour, but this extra weight restricts movement and limits the animal's maximum possible size. It is only in water, where body weight is countered by buoyancy, that crustaceans such as the American lobster can grow into real giants.

### BUOYED BY WATER

Weight – determined by the force of gravity and measured in Newtons – changes according to the medium in which an object exists (air or water). Mass – or the amount of material, measured in kilograms – is fixed. In the ocean, where gravity's pull is partly offset by upthrust from the water, a large lobster has enough muscle power to move its limbs and walk on the seabed. But on land, the lobster's weight is much greater, rendering it practically immobile.

Weight of lobster in air: 9.8 Newtons

Mass of lobster is 1 kg

Weight of lobster in water: 6 Newtons

Mass of lobster is still 1 kg

**WEIGHT AND MASS OF LOBSTER IN AIR AND WATER**

### Replaceable armour
Native to the northwest Atlantic, the American lobster (*Homarus americanus*) is the world's heaviest arthropod: the largest recorded specimen exceeded 20 kg (44 lb), although most of its kind weigh much less. Like all arthropods, it must periodically moult its exoskeleton and build new armour to get bigger. Since the American lobster grows throughout its life, the record-breaker might have been over a century old.

**Long antennae** are packed with tactile receptors for sensing in dark or murky waters

**Pincerlike chelae** on the first pair of legs are enlarged for defence against predators

**Carapace** is part of the exoskeleton that forms a shield covering the head and leg-bearing thorax, which are fused together

**Elongated abdomen** is divided into six articulated segments, which allows the body to flex

**Five pairs of walking legs** have multiple joints and are tipped with small chelae

**Tail fan** acts like a paddle when the abdomen is flexed, propelling the lobster backwards – and away from danger

**Cirri** are strong enough to support the weight of the feather star; they can move back and forth to help the animal "walk" over the seabed

**Arms pointing upwards** can better catch food particles drifting down from above

## Orifices upwards

The mouth at the centre of a feather star faces upwards – not downwards, as in starfish – so it is better positioned to receive food particles collected by the arms. Food passes through a U-shaped gut, with undigested waste emerging from an anus, which is also upward-facing.

Calyx (central body)
carries the feather star's
upward-facing mouth
and contains a short,
coiled gut

# suspension feeding

Animals that are nourished by tiny food particles suspended in sea water do better if they can spread their body over a wide area to catch more particles. Feather stars – relatives of starfish – do just that, directing feathery arms into incoming currents to trap plankton and floating fragments of detritus, while they "walk" along the seabed on spindly, limblike cirri. Some feather stars can also sweep their arms up and down to swim in mid-water.

Pinnules of each arm fan
outwards in the water
to help trap food

**Feathery arms**
Feather stars have arms that grow
in multiples of five. Some species
have as many as 200 arms. Each
arm is equipped with myriad side
branches, called pinnules.

## COLLECTING PARTICLES

True filter-feeders, such as clams, pump sea water through a sieve-like part of the body, but suspension-feeders, such as feather stars, effectively "comb" the water to find food. Hydraulically operated tube feet along a feather star's arms flick food into grooves, where microscopic, hairlike structures called cilia sweep it towards the mouth.

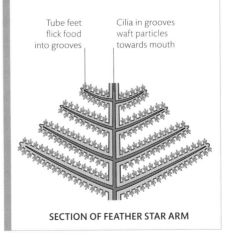

Tube feet
flick food
into grooves

Cilia in grooves
waft particles
towards mouth

**SECTION OF FEATHER STAR ARM**

**Under the Great Wave off Kanagawa (1829–33)**
The fierce energy of Hokusai's huge wave, poised
above the boats caught in its grasp, has made this
work (from Hokusai's series *36 Views of Mount Fuji*)
the most recognized of all Japanese prints. Although
it was common for prints at the time to feature a
variety of colours, here a limited palette was used to
powerful effect to emphasize the wave's dominance.

*the ocean in art*

# the great wave

Since early times, artists in Japan – an island nation surrounded by the crashing waves of the Pacific and the complex currents and whirlpools of the Sea of Japan – have sought inspiration in the sea. In the Shinto faith, everything in nature, from the rocky shore to the ocean depths, has its own life force: it is this energy that is almost tangible in Japanese woodcuts.

Woodblock printing, the method used to create woodcuts, was introduced to Japan from China around 2,000 years ago, but it was not until the Edo period (1603–1868) that the art form was popularized. While only the wealthy could afford paintings, woodcuts were available to all, being printed in their thousands and costing as little as a bowl of noodles. The prints were a collaboration between an artist, who drew the design on paper, a cutter, who carved the design onto wooden blocks through the paper, and the publisher of the print runs. Early prints were monochrome, but by the mid-18th century colour prints, known as "brocade pictures", were created using separate blocks for up to 20 colours.

Traditionally, the woodcuts depicted city scenes, including geisha, courtesans, and Kabuki actors, and erotica. They became known as *ukiyo-e*, meaning "pictures of the floating world" – a reference to transitory pleasures. However, towards the end of the Edo period there was an increasing interest in the natural world, and woodcut artists responded by creating landscapes and seascapes. They drew on literary traditions, while engaging wholly with the moods and rhythms of the sea.

Among the most enduring of these woodcut artists are Hokusai, best known for *Under the Great Wave off Kanagawa*, and Hiroshige, often described as Japan's last great *ukiyo-e* master. Hiroshige's woodcuts reflect the sea in every guise, from waves crashing on rocks and a tranquil distant shore in *The Sea off Satta Peak, Suruga Province* (1858) to the treacherous currents of *Naruto Whirlpools, Awa Province* (1855).

After Japan opened up to the West in the 1850s, following centuries of isolation, such prints were avidly sought in Europe and the US and revered by Impressionist and Post-impressionist artists alike.

> ❝ These waves are claws, the boat is caught in them, you can feel it. ❞

LETTER FROM VINCENT VAN GOGH TO HIS BROTHER THEO, 1888

Hard jaws are made up of modified ossicles

**Brittle star mouth**
The tube feet on a brittle star's arms pass food to tiny jaws flanking the mouth beneath the central disc. The food enters a short stomach; undigested waste passes out through the mouth, since the gut has no separate anus.

Muscular arms bend in multiple directions to enable movement across the uneven seabed

Mucus-covered tube feet select particles of food and bundle them together for passage towards the mouth

# walking with arms

The thick, reinforced skin of a starfish undermines its flexibility, so the animal must crawl slowly using hundreds of tiny, suckerlike "tube feet" (see pp.220–01). But in brittle stars – relatives of starfish – the skin is more supple, and a complex lattice of muscles allows their arms to move more freely in the horizontal direction. By snaking to and fro, the spiny arms of a brittle star can grip the substrate and push forwards; rather than being involved in locomotion, the suckerless tube feet are free to gather food particles.

### BODY SUPPORTS
Like other echinoderms (urchins, starfish, feather stars, and sea cucumbers), a brittle star has a body that is supported both by hard skeletal ossicles that reinforce the skin and by a system of canals circulating sea water (a water-vascular system) that transports materials. Muscles running along the length of the arms control the arms' movements.

Ossicles are hard plates in the skin

Longitudinal muscles contract to bend the arm

Water-filled channels move the tube feet

**CROSS SECTION OF BRITTLE STAR ARM**

**Brittle arms**
A European common brittle star (*Ophiothrix fragilis*) uses its sinuous arms to walk over the seabed while the animal grazes on detritus and microbes. As the brittle star's name suggests, the arms break easily, although they can regenerate after injury.

**Arms** radiate
from centre
(typically 8–15
in common sun star)

**Tube feet** have
powerful suckers

## Sticky ends

Each tube foot on a starfish ends in a disc, which secretes adhesive and de-adhesive chemicals. These allow the starfish to bind temporarily to other surfaces.

Tube feet emerge from grooves along each arm

# tube feet

One of the traits that define the group of marine animals known as echinoderms is the possession of multifunctional structures called tube feet. They are the main organs of locomotion in starfish and urchins, they sometimes act as touch-sensitive feelers, and their thin skin serves as a surface for gas exchange. They also play a major role in the feeding process, helping to pass tiny particles to the mouth, and in starfish their adhesive strength is used to prise open the shells of prey such as mussels and clams.

Bunched spines aid grip on seabed

Tip of arm contains tiny, light-sensitive eyespots

## On the move

The action of tube feet allows the common sun star (*Crossaster papposus*) to creep across horizontal, vertical, or overhanging surfaces. Each arm contains sensory structures, such as eyespots, which enable the starfish to tell light from shade.

### THE WATER-VASCULAR SYSTEM

Tube feet are part of a system that uses water pressure to aid locomotion, feeding, and respiration in echinoderms. In starfish, water enters (and leaves) the system via a pore on the top of the body. It then circulates to the ring canal and into the radial canals, which extend along each arm. From there the water can pass into the tube feet, each of which comprises a bulbous sac (ampulla) and a stretchy foot (podium). Contraction of the ampulla forces water into the foot, making it extend.

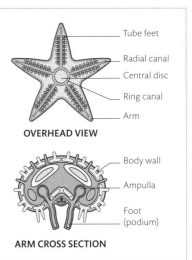

Tube feet
Radial canal
Central disc
Ring canal
Arm

**OVERHEAD VIEW**

Body wall
Ampulla
Foot (podium)

**ARM CROSS SECTION**

## Number of arms

Most starfish species have five arms, but some have 10, 20, or even 50. The underside of each arm is covered in thousands of tiny tube feet, which enable the animal to "walk" over rocks, "climb" seaweed, and grab its prey. Each arm contains a full set of the starfish's body systems, so if the animal loses a limb it can survive without it. A starfish can also regenerate a lost arm.

Short projections radiate from cushionlike central disc

**CUSHION STARFISH**
Cushion star
*Ceramaster granularis*

Upper body is covered in small "bony" plates

**FIVE-ARMED STARFISH**
Cuming's sea star
*Neoferdina cumingi*

Long, thin arms can be 12–15 cm (5–6 in) long

**6- TO 16-ARMED STARFISH**
Nine-armed sea star
*Luidia senegalensis*

## Size

The diameter of a starfish can range from a few millimetres to 1 m (3 ft). The size of the central disc can also vary. Some, as in the Pacific blood star (*Henricia leviuscula*), are very small relative to the arm length; others are very large. The largest known species – the sunflower sea star (*Pycnopodia helianthoides*) – can weigh 5 kg (11 lb) and live for 35 years.

Arm is edged with paddlelike spines

**0.5–1 CM (⅕–⅓ IN )**
Paddle-spined sea star
*Allostichaster palmula*

Central disc is less than one-fifth of the total diameter

**8–12 CM (3¼–5 IN)**
Pacific blood star
*Henricia leviuscula*

Stubby arms are covered with brightly coloured spines

**UP TO 30 CM (1 FT)**
Panamic cushion star
*Pentaceraster cumingi*

## Habitat and depth

Starfish can be found in every type of marine environment, from the icy polar oceans to tropical habitats. Some permanently inhabit shallow waters, living in tidal pools and on rocky shores, while others live on kelp beds and all parts of the coral reefs. Species have been observed at depths of 9,000 m (30,000 ft).

Smooth upper surface covered with layer of mucus

**COASTAL TO 100 M (330 FT)**
Leather star
*Dermasterias imbricata*

Colour is derived mainly from the blue skin pigment linckiacyanin

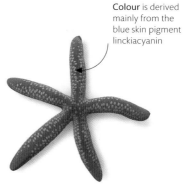

**SURFACE TO 70 M (230 FT)**
Blue linckia
*Linckia laevigata*

Entire surface is covered with small spines called pseudopaxillae

**100–1,000 M (330–3,300 FT)**
Common sunstar
*Crossaster papposus*

**16- TO 25- ARMED STARFISH**
Crown-of-thorns starfish
*Acanthaster planci*

# starfish

A group of more than 2,000 species of mostly predatory marine invertebrates, starfish (or sea stars) are echinoderms, closely related to sea cucumbers and sea urchins. Found in all the world's oceans, they are recognized for their – usually five-point – radial symmetry. They have no brain or blood. Instead a water-vascular system moves sea water through their bodies, delivering the key nutrients needed for their organs to function.

**Underside** of body has more than 15,000 tube feet with suction pads

**UP TO 1 M (3 FT)**
Sunflower sea star
*Pycnopodia helianthoides*

**Jewel of the sea**
The honeycomb cushion star (*Pentaceraster alveolatus*) is found in the intertidal regions and reef platforms of the Indo-Pacific at depths of 1–60 m (3–200 ft). Up to 40 cm (16 in) across, this starfish lives singly and in groups close to seagrass beds and a supply of microalgae.

**Rows of prominent** spines along centre and sides of each arm

**Light-sensitive cells** at the end of each arm allows animal to navigate, hunt for food, and hide from predators

**Cushionlike** central disc and arms

**Sieve plate** (madreporite) where water enters water-vascular system

**MORE THAN 1,000 M (3,300 FT)**
Deepsea star
*Porcellanaster ceruleus*

**Five-rayed symmetry**
Most echinoderms have five axes or arms radiating from the centre. On this collector urchin (*Tripneustes gratilla*), these are seen as five double rows of tube feet (see p.221), interspersed with bands of mostly orange spines.

Spines and tube feet are arranged in alternate bands radiating from the mouth

# spiny skin

Echinoderms are a large and highly diverse group of invertebrates found only in the sea. Their name means spiny skin, with the spiniest in the group being the sea urchins. The bodies of sea urchins are covered in long, mobile spines, which can rotate at the base and are used for manoeuvring and to repel predators or encrusting animals. The spines vary according to the species, ranging from fine and bristly to stout and pencil-like.

**Tube feet**
can extend well beyond the body

Hole at top is exit point for digestive and reproductive tracts and the water-vascular system (see p.221)

Pore was once the emergence point for a tube foot

**Urchin test**
A sea urchin's body is supported by a shell-like skeleton called a test. When the urchin is alive, the test is covered in a thin layer of soft tissue, spines are attached to rounded protuberances (tubercles) via ball-and-socket joints, and tube feet extend from the tiny holes (pores). When the urchin dies, the spines fall off and the tube feet and tissue decay, leaving only the test.

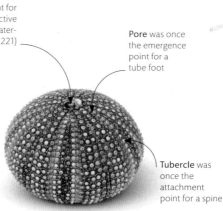

Tubercle was once the attachment point for a spine

## CLEANING AND STINGING

Between the spines covering a sea urchin's body are tiny, claw-shaped appendages mounted on a flexible stalk. Each appendage (pedicellaria) consists of three movable jaws (valves). Some are used as pincers, to pick debris or algae from the urchin's body, while others are connected to venom glands, enabling the urchin to deliver a stinging nip.

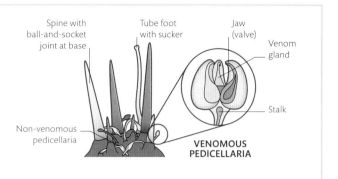

Spine with ball-and-socket joint at base

Tube foot with sucker

Jaw (valve)

Venom gland

Stalk

Non-venomous pedicellaria

**VENOMOUS PEDICELLARIA**

**Spines** can swivel and be raised or lowered, useful for wedging the urchin into a crevice or lifting it slightly off the seabed

# scalloped hammerhead shark

All nine species of hammerheads are instantly recognizable by their wide, flattened heads, known as cephalofoils. Despite its appearance and size, the scalloped hammerhead (*Sphyrna lewini*) is generally not aggressive to humans, but our activity has put the iconic animal on the endangered list.

Also known as the kidney-headed shark, the scalloped hammerhead is named for the notched front edge of its head. Its eyes and nostrils are positioned on each end of the "hammer", giving it better binocular vision and a heightened sense of smell. The cephalofoil may also provide a larger surface area for the shark's sensory cells, or ampullae of Lorenzini, which enable the animal to sense electrical fields given off by its prey. Scalloped hammerheads have relatively small mouths so live on a diet of fish and invertebrates they can swallow whole. They have been observed using one side of the cephalofoil to pin down a stingray before eating it.

Adult scalloped hammerheads grow to around 3.6–4 m (12–13 ft) long. They are found in most temperate and tropical oceans, usually at depths of 25–275 m (80–900 ft). They often swim alone or in pairs, but periodically huge numbers of adults form schools near underwater sea mounts or off island coasts.

Mature animals give birth to large litters of live young (anything from 12 to 41 pups) in coastal waters, but many are eaten by other sharks. Fishing and the shark-fin trade have also taken their toll on juvenile and adult animals.

### Shark super school
Scalloped hammerheads gather in their hundreds to form huge schools. Theories as to why they do this vary from migration to aggression and sexual displays, but it makes them highly vulnerable to targeted fishing.

**Front teeth** are small and serrated; larger, flatter back teeth are used to grind up shellfish

**Eye** at each end of cephalofoil gives shark 360-degree field of vision

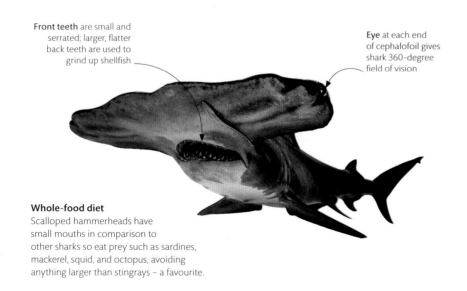

### Whole-food diet
Scalloped hammerheads have small mouths in comparison to other sharks so eat prey such as sardines, mackerel, squid, and octopus, avoiding anything larger than stingrays – a favourite.

**Taking to the air**
Large groups of mobula rays sometimes perform spectacular breaching displays, leaping high into the air. This behaviour is thought to be a form of communication.

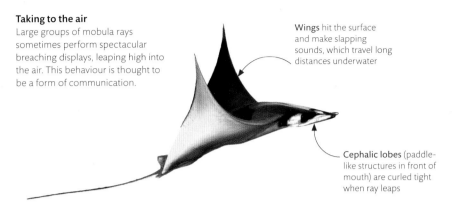

**Wings** hit the surface and make slapping sounds, which travel long distances underwater

**Cephalic lobes** (paddle-like structures in front of mouth) are curled tight when ray leaps

# flying under water

Skates and rays, which are very similar-looking cartilaginous fish, have evolved a unique form and swimming style, using their flattened body and greatly enlarged pectoral fins. Many species are bottom-dwellers and ripple their fins to generate propulsion over the seabed. However, the closely related mobula and manta rays, both of which venture higher in the water column to filter-feed, beat their fins up and down – a mode of swimming that resembles the flight of birds and bats.

**Perpetual motion**
Mobula rays swim continually to maintain a flow of water over their gills. They travel widely, often swimming together in large groups as here, off the coast of Baja California, Mexico. The smaller mobulas are particularly gregarious, frequently gathering in schools of thousands.

## FEEDING ON THE WING

When feeding, mobula and manta rays swim slowly, using the motion to generate a feeding current. They are filter-feeders, with a mouth that opens to the front, rather than beneath as in other rays. A pair of cephalic lobes unfurls to funnel sea water and plankton into the mouth. Inside, the feeding current runs over the gill rakers, causing the plankton to ricochet off the filter and flow on to the throat, while the water exits via the gills.

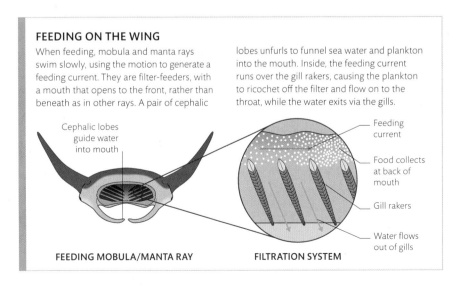

Cephalic lobes guide water into mouth

Feeding current

Food collects at back of mouth

Gill rakers

Water flows out of gills

**FEEDING MOBULA/MANTA RAY**　　　**FILTRATION SYSTEM**

**When hunting,** barracudas often swim in schools and head straight towards their prey, relying on short bursts of speed

**Careful adjustment**
Chevron barracudas (*Sphyraena putnamae*) are
fast-swimming predators. They have a large
swim bladder, which regulates their buoyancy
and saves energy. Adjusting the volume of
gas in the swim bladder to compensate for
pressure changes when rising or descending
can slow the fish down. As a result, barracudas
generally make only short vertical movements.

Torpedo-shaped,
streamlined body
is packed with muscle,
making barracudas swift
and powerful swimmers

# swim bladders

As every scuba diver knows, maintaining a steady position in the
water can be hard work. For most bony fish living in open water,
the energy-saving solution is a gas-filled swim bladder, careful
control of which gives the fish neutral buoyancy and an ability to
hang effortlessly at a certain depth in the water, without floating
upwards or sinking. Bottom-dwelling fish often lack a swim
bladder, and the feature is absent from sharks and rays, which rely
instead on constant swimming and have a large liver filled with
low-density oil, called squalene, which increases their buoyancy.

## ADJUSTING BUOYANCY

The swim bladder is a thin-walled sac, which can expand and contract depending
on how much gas (mostly oxygen) is inside. When a fish swims upwards, the water
pressure drops, which makes the gas in its swim bladder expand. To stop the fish
bobbing to the surface, gas is removed from the bladder. The reverse happens when
the fish swims downwards: increased pressure means gas is added to prevent sinking.

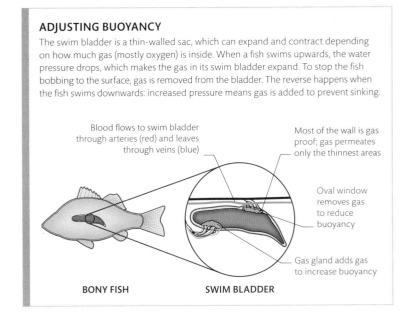

Blood flows to swim bladder
through arteries (red) and leaves
through veins (blue)

Most of the wall is gas
proof; gas permeates
only the thinnest areas

Oval window
removes gas
to reduce
buoyancy

Gas gland adds gas
to increase buoyancy

**BONY FISH**          **SWIM BLADDER**

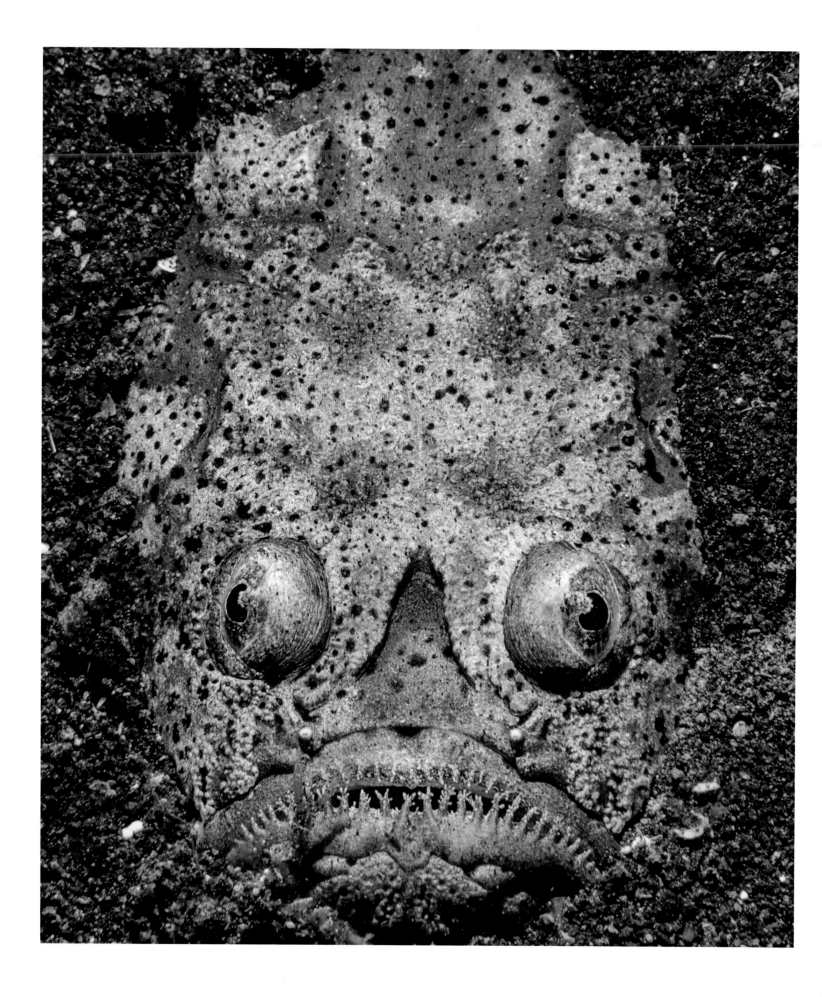

# keeping out of sight

Crypsis is the strategy of remaining hidden. One component of it is camouflage – blending against a background, through body coloration, pattern, and texture – but behaviour is also a key factor. In stargazers, the purpose of invisibility is both offensive and defensive; achieving it means remaining partially buried in sediment on the seafloor, where the fish waits for a chance to ambush prey that ventures within striking distance.

### ATTRACTING PREY

Stargazers tempt prey within range using a scrap of flesh, known as a lure, protruding from the mouth. The lure resembles a small, soft-bodied invertebrate, such as a ragworm, and it can be twitched invitingly to gain the attention of potential prey. Bulging eyes watch intently, and when a smaller fish approaches closely enough, the stargazer opens its huge, upward-facing mouth in a few hundredths of a second, creating a surge of inrushing water that the prey is powerless to resist.

**Lying in wait**

The whitemargin stargazer (*Uranoscopus sulphureus*) lies partially hidden in the sand of the seabed. Patchy colours and granular skin texture blend with the sand, and by shuffling its pectoral fins and tail under the sediment, the stargazer disrupts outlines and shadows that might alert prey to danger. Sometimes only the swivelling eyes protrude.

232 • 233

coastal seas

**Shock treatment**

Like all stargazers, the Atlantic stargazer (*Uranoscopus scaber*) has cryptic coloration. It is also one of the few marine species (together with the whitemargin stargazer) capable of bioelectrogenesis – the ability to produce a jolt of electricity to deter predators.

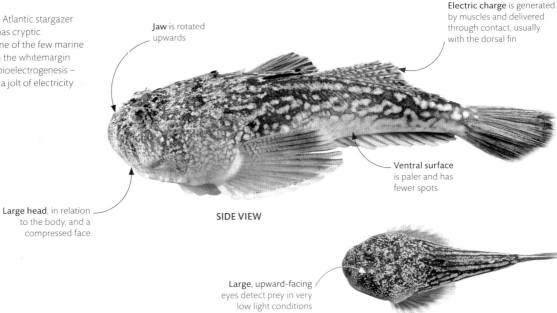

Jaw is rotated upwards

Electric charge is generated by muscles and delivered through contact, usually with the dorsal fin

Ventral surface is paler and has fewer spots

Large head, in relation to the body, and a compressed face

**SIDE VIEW**

Large, upward-facing eyes detect prey in very low light conditions

**OVERHEAD VIEW**

**Uplifting coastline**
The scenic shores of the Lofoten Islands, an archipelago off the northwestern coast of Norway, provide an example of a coastline that has been uplifted by tectonic activity.

# sea-level change

Sea level is the average height of the ocean surface relative to land. This has changed throughout Earth's history in response to climate change and the movement of tectonic plates. Eighty million years ago, as a result of an expansion of mid-ocean ridges (see pp. 264–65) that caused displacement of water upwards, sea level was 250–350 m (820–1,150 ft) higher than today. At the peak of the last Ice Age (about 25,000 years ago), when a huge volume of water was stored on land in polar ice caps, sea level was 120 m (400 ft) lower than it is today. As global warming melts the ice caps and warmer sea water expands, sea level is once again on the rise — perhaps by as much as 50 cm (20 in) by the end of the 21st century.

## REGIONAL SHORELINE CHANGES

Relative sea level may rise or fall in areas of tectonic activity that either uplifts the landmass or causes it to sink. Where the landmass rises, the relative sea level drops and the shoreline advances. Conversely, where the landmass sinks, the relative sea level rises and the shoreline retreats. Both processes produce localized changes to the shoreline, but do not have an overall impact on global sea level.

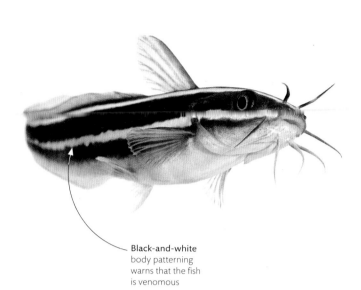

**Black-and-white** body patterning warns that the fish is venomous

**Detecting movement**

The striped eel catfish (*Plotosus lineatus*) – the only catfish found in coral reefs – has highly developed sensory organs, including the lateral line, which comprises a series of pores and canals along its flanks. Sensory cells in the lateral line detect vibrations (including low-frequency sound waves) and pressure changes, helping the nocturnal fish to navigate dark water and sense the movement of prey it cannot see.

# all-over senses

Water is an excellent medium for transmitting a wide range of sensory information, including vibrations and electrical signals. Many fish possess well-developed senses, with those of the catfish family being particularly acute. Their large eyes are useful because light transmission is limited in water, and their hearing is enhanced by tiny bones that transmit sound waves picked up by the swim bladder to the inner ear. In addition, catfish have chemical and electrical receptors all over the surface of their body.

## CHEMORECEPTOR ORGANS

Catfish are equipped with between 250,000 and several million chemoreceptor organs, which are similar to taste buds but located all over the body, in differing degrees of density. These allow the fish to taste anything they come in contact with, enabling them to hunt effectively in the dark. The taste buds do not have to touch potential food in order to taste it. Any soluble chemicals that come from potential food can diffuse through the water to reach the taste buds and stimulate them.

High density of taste buds along leading edge of the fin

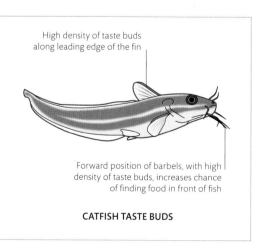

Forward position of barbels, with high density of taste buds, increases chance of finding food in front of fish

**CATFISH TASTE BUDS**

**Elongated dorsal fin** merges with tail fin, giving eel-like appearance

**Lateral line** detects vibrations and pressure changes

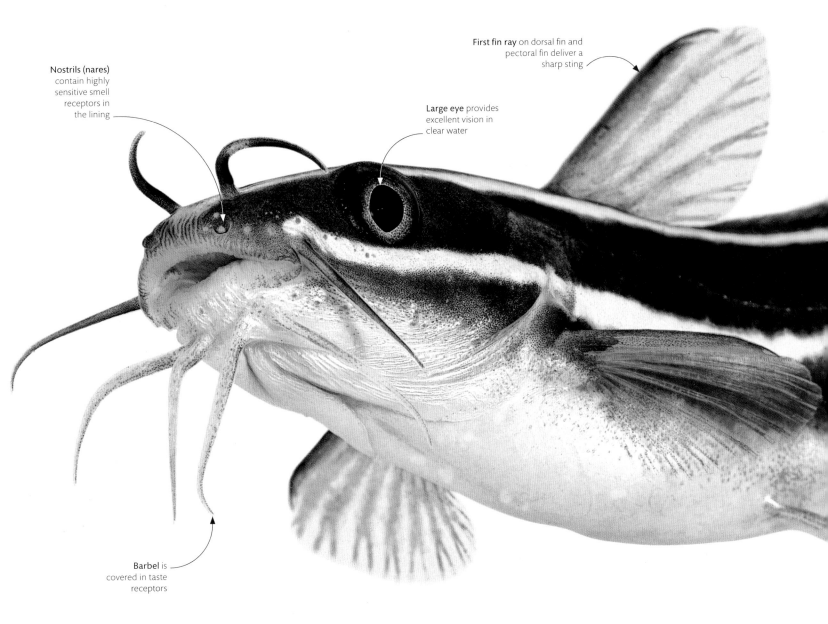

**Nostrils (nares)** contain highly sensitive smell receptors in the lining

**Large eye** provides excellent vision in clear water

**First fin ray** on dorsal fin and pectoral fin deliver a sharp sting

**Barbel** is covered in taste receptors

## Sensitive body
Pressure-sensitive pores all over a sardine's body enable it to detect the movements of other fish nearby and react instantly, for example by forming a tight group.

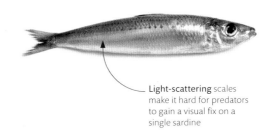

**Light-scattering** scales make it hard for predators to gain a visual fix on a single sardine

# migrating in schools

Between spring and summer, an explosion of plankton growth occurs in nutrient-rich temperate seas and many fish arrive from warmer areas to feed. Among these are sardines and anchovies, both small fish in the herring family, which migrate in vast schools of millions of individual fish, following seasonal currents. During these migrations, these immense schools draw vast numbers of predators, including seabirds, sharks, dolphins, and whales, into so-called feeding frenzies.

## Safety in numbers
When a sardine school (in this case *Sardina pilchardus*) senses a threat, it forms an even tighter group, with each fish attempting to stay in the centre of the crowd, where it is less likely to be picked out. The swirling mass of bodies shifts and changes in shape in an attempt to confuse predators.

---

### GROUPS OF FISH

Fish often gather in small or large groups. A shoal is a relatively casual aggregation of fish, which move in different directions and include individuals of one or more species. A school is a much more organized group, in which all members move in a coordinated manner. Schooling behaviour almost always has a defensive function, and smaller fish, such as those in the herring family, often rely on swimming in close-knit schools for their survival.

Loose group of individual fish facing in different directions

**SHOAL**

Tight group moves as one in a single direction

**SCHOOL**

# great frigatebird

Found on tropical islands in the western Atlantic, Pacific, and Indian oceans, the great frigatebird (*Fregata minor*) soars above the waves for days, sometimes weeks, at a time. It is nicknamed the "pirate bird" due to its habit of aggressively stealing food from the beaks of other seabirds.

The great frigatebird feeds mainly on flying fish and squid and flies for hours, watching for prey to break the surface, and then seizing it with its long, hooked beak. While it acquires most of its food this way, it is also a kleptoparasite – an animal that steals prey caught by another animal. Frigatebirds attack other seabirds, especially boobies, in mid-air as they return to their nests, harassing them into giving up meals intended for their chicks. This "mugging" of other birds has earned frigatebirds their common name, as their behaviour resembles that of the man-o'-war ships, or frigates, used by pirates on ocean raids (and they are sometimes also called man-o'-war birds).

At up to 105 cm (41 in) long and weighing 1.5 kg (3 lb 4 oz), but with a wingspan of up to 2.3 m (7 ft 6 in), the great frigatebird has the largest ratio of wing area to body mass of any living bird. The area of the wing, as well as its long, narrow shape, enables the bird to fly vast distances in search of food, although it usually stays within 160 km (100 miles) of land in order to roost.

However, this aerial specialist has a significant weakness – its feathers are not waterproof, as they lack sufficient preen oil to protect them from sea water. A frigatebird never willingly gets wet as, if its plumage becomes waterlogged, it may be unable to take off again.

### Attracting a mate
Like the related cormorants and boobies, frigatebirds have a throat pouch - called a gular sac. A male inflates his huge, bright-red pouch to attract a mate. The females also have gular sacs but never inflate them.

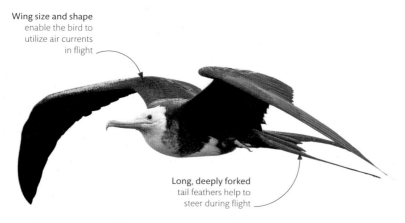

**Wing size and shape** enable the bird to utilize air currents in flight

**Long, deeply forked** tail feathers help to steer during flight

### Long-distance soaring
All birds have air-filled bones to make them lightweight, but a frigatebird skeleton forms a smaller fraction of its body weight than any other bird, making it the nimblest flyer. It can soar for weeks at a time and sleep mid-flight.

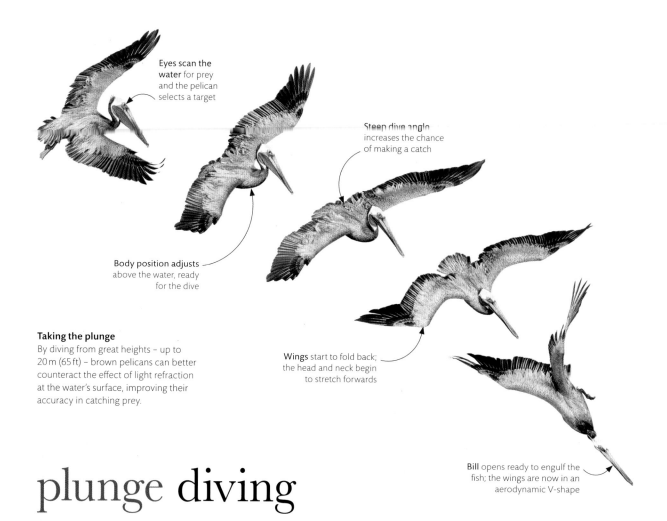

Eyes scan the water for prey and the pelican selects a target

Steep dive angle increases the chance of making a catch

Body position adjusts above the water, ready for the dive

**Taking the plunge**
By diving from great heights – up to 20 m (65 ft) – brown pelicans can better counteract the effect of light refraction at the water's surface, improving their accuracy in catching prey.

Wings start to fold back; the head and neck begin to stretch forwards

Bill opens ready to engulf the fish; the wings are now in an aerodynamic V-shape

# plunge diving

Birds that catch fish underwater must overcome the buoyancy of bodies adapted for flying. But by plunging into the water from great heights, diving specialists such as gannets can reach deeper prey before floating back upwards. Brown pelicans do likewise to take advantage of vast shoals in coastal seas – a strategy that sets them apart from freshwater pelicans, which scoop up their catch as they paddle over the surface.

**KLEPTOPARASITES**
Because pelicans spend time draining water from their pouch and manipulating their catch before swallowing it, these birds are vulnerable to thieves, or kleptoparasites. In the Caribbean, laughing gulls (*Larus atricilla*) often land on the head of juvenile pelicans because the young birds are more likely to spill their fish. But in the Gulf of Mexico, Heermann's gulls (*L. heermanni*) harass adults that have a bigger catch.

**ADULT AND JUVENILE LAUGHING GULL (*LARUS ATRICILLA*)**

Skin pouch hanging from the bill's lower mandible expands to accommodate several litres of fish-filled water

**Grabbing the catch**
The brown pelican targets a single fish but usually scoops up several in its pouch – together with a heavy volume of water, which must be drained before the catch can be swallowed.

**Coastal pelican**
The brown pelican (*Pelecanus occidentalis*) of Central America and the Caribbean fishes along shallow inshore coastal seas. Here, upwellings of nutrient-rich water support some of the biggest shoals of anchovies and sardines on the planet. The bird's chocolate-brown body contrasts with the largely white plumage of its freshwater cousins.

**Upper mandible**
of the pelican's bill is reinforced with a strong keel, which helps it support the weight of a heavy catch

**Long neck** enables the pelican to rest its heavy bill against its breast when on the ground or flying

**Dark brown plumage** is waterproofed with an oily coating

**Tip of the bill** has a sharp, ridged "nail" that can grip slippery fish firmly

**Using tools**
Back on the surface after a dive, an otter places a carefully chosen rock on its front, then bashes a clam against it to crack open the shell and access its meal.

*spotlight species*

# sea otter

Superbly adapted to its coastal habitat, the sea otter (*Enhydra lutris*) has the densest coat of any animal on Earth, powerful webbed hind feet, a strong, rudderlike tail, and heightened senses of smell and touch that help it to locate prey in murky waters. It is also one of the few mammals to use tools.

Found in the north Pacific Ocean, the sea otter is a member of the Mustelidae family, a group that includes weasels and badgers. One of the smallest marine mammals, an adult sea otter has a body length of around 1.2 m (4 ft). Its body is streamlined, highly buoyant, and – apart from its pads and nose – entirely covered in a dense, two-layered coat of fur. Short, insulating underfur contains up to 155,000 hairs per sq cm (1 million per sq in), while an outer layer of longer guard hairs forms a waterproof barrier against cold seas. Grooming helps the sea otter to introduce insulating air bubbles into its fur. Unlike seals or sea lions, otters lack a blubber layer so are dependent on their fur to keep warm and dry.

A sea otter spends almost all its life at sea. Besides mating and giving birth in water, otters float on their backs on the surface to eat, sleep, and groom. They dive up to 75 m (250 ft) to forage on the seabed, closing their nostrils and ears while underwater. A substantial food supply is crucial – adult sea otters must eat 20–33 per cent of their body weight each day to stay alive. Long whiskers and sensitive forepaws help detect vibrations in low-visibility conditions, allowing these animals to find clams, sea urchins, crabs, and other invertebrates in near-darkness. Sea otters may stash several prey items in pouches of loose skin beneath their forearms, enabling them to carry more food back to the surface.

**Hanging on**
Sea otters often form groups known as "rafts" when resting on the surface. As well as wrapping themselves securely in kelp, some otters prevent themselves from drifting away by holding another animal's forepaws.

**Ears** are small and lack external parts (pinnae), but are very sensitive; dugongs rely more on hearing than sight

**Stout whiskers** sense objects on seabed, even in cloudy water

**Small eyes** high on head provide good field of view, although eyesight is relatively poor

**Wide, fleshy oral disc** can be pushed forwards when sensing or manipulating food

**Muscular lips** uproot plants from sediment

**Grazing pair**
Dugongs spend most of their time alone or in pairs. Their feeding habits (known as cultivation grazing) look destructive, but their regular visits result in regrowth of some seagrass varieties, which would otherwise be crowded out by tougher, more fibrous plants.

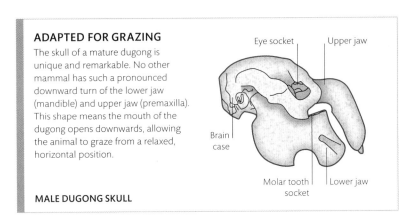

**ADAPTED FOR GRAZING**
The skull of a mature dugong is unique and remarkable. No other mammal has such a pronounced downward turn of the lower jaw (mandible) and upper jaw (premaxilla). This shape means the mouth of the dugong opens downwards, allowing the animal to graze from a relaxed, horizontal position.

Eye socket    Upper jaw

Brain case

Molar tooth socket    Lower jaw

**MALE DUGONG SKULL**

# grazing on grass

Sirenians are a group of unusual sea mammals that include the Atlantic manatees and the Indo-Pacific dugong (*Dugong dugon*). Collectively, they are sometimes known as sea cows because of their grazing lifestyle. They are almost wholly herbivorous, favouring tender seagrasses and algae. Since they have a low-energy diet, dugongs take life at a leisurely pace and are restricted to warmer tropical and subtropical waters, where they do not need to burn energy to maintain a constant warm body temperature.

**Grazing trail**
Dugongs eat every part of their food plant, including the roots, so they create meandering trails of bare sediment as they wander through seagrass beds.

**Golden trevally** take advantage of small invertebrates dislodged by the dugong's feeding activity

# humpback whale

While humpback whales (*Megaptera novaeangliae*) are remarkable in many ways – their size, intelligence, and the distances they swim each year – these oceanic mammals are particularly well known for their sophisticated vocalizations, or "songs", produced by males during mating season.

Humpbacks get their common name because of the way they arch their backs when diving, creating a humped profile. At up to 60 m (200 ft) long, they are not the largest whales in the sea, but they do hold the record for the longest pectoral fins, which can reach up to 5 m (16 ft) – up to a third of their total body length. Mothers and calves have been observed touching fins while swimming, possibly as a sign of reassurance.

Male humpbacks vocalize for up to 20 minutes at a time, creating songs that span several octaves and may be heard up to 30 km (20 miles) away. All males in a group sing the same combination of squeaks, groans, howls, and moans, sometimes adding subtle variations but keeping the theme broadly the same. However, studies show that these songs evolve, developing new themes every few years.

Humpback whales make one of the furthest migrations of any mammal on Earth. Many undertake annual round-trips of around 16,100 km (10,000 miles), spending summer in feeding grounds rich in krill and other plankton – their main source of food – before migrating to spend winter in the warmer waters of their breeding sites. Humpbacks were pushed to near-extinction in the 20th century. However, conservation efforts have ensured that most of the recognized populations are no longer endangered.

### Marks of distinction

All humpbacks have dark upper bodies, but white underpart patterns, found on bellies and beneath tail flukes as well as pectoral fins, are unique to individuals. Whales from the southern hemisphere have whiter undersides than northern-hemisphere whales.

## GROUP HUNTING

Humpbacks often cooperate to hunt, using various sounds and fin slaps to herd or disorient their prey. "Bubble-netting" is unique to humpbacks: the whales use their blowholes to create curtains of air bubbles, corralling prey into a dense mass, before exhaling more bubbles underneath to drive them upwards. They then swim open-mouthed up through the centre of the net, capturing mouthfuls of prey.

Whales dive to bottom of curtain

Whales swim through net with mouths agape

Rising bubbles form cylindrical corral

Whales swim around prey, exhaling bubble curtain

**BUBBLE-NET FEEDING**

**Rich waters**
Off the coast of California, USA, cool currents
from the Arctic combine with offshore winds
to bring nutrients to the surface, leading to the
growth of phytoplankton, shown here in green.

# upwelling and downwelling

Vertical currents below the ocean surface bring water up towards the surface (upwelling) or carry water downwards (downwelling). These currents are most common in coastal regions, but can also occur far from land. The quiet central parts of ocean gyres (surface current circulation systems) are areas where deepwater currents converge and downwelling results. Extreme downwelling occurs in areas of the Norwegian-Greenland Sea and around Antarctica, where ice-cold dense water sinks and is carried south by bottom currents. When upwelling occurs, the rising current brings nutrients nearer to the surface, supporting plankton growth, which in turn attracts a wide variety of ocean life. Major upwelling zones exist off Peru, Namibia, and western Canada.

## THE EFFECTS OF CURRENTS AND WINDS

Coastal upwelling of water occurs where surface currents are deflected offshore due to the effect of Earth's rotation combined with that of the prevailing winds. The movement of surface water away from the shore draws water upwards from the ocean depths. In regions where currents near the surface are deflected towards the coast, the waters are forced downwards to produce downwelling.

Wind direction

Surface waters move
away from coast

Wind
direction

UPWELLING

Water drawn
upwards

Surface waters
move towards
coast

Water forced
downwards

DOWNWELLING

# atlantic spotted dolphin

Inquisitive, acrobatic, playful, and highly sociable marine mammals, Atlantic spotted dolphins (*Stenella frontalis*) are found in temperate to tropical Atlantic waters from southern Brazil to New England via the Gulf of Mexico in the west, and from Angola to Morocco in the east.

Although named for their distinctive mottled skin, these dolphins are born spotless and do not acquire any speckling until they are eight to 15 years old; some remain unspotted so are misidentified as slender bottlenose dolphins. Calves are 60–120 cm (2–4 ft) at birth, but adults can be 1.7–2.3 m (5 ft 6 in–7 ft 6 in) long and weigh 110–141 kg (240–310 lb).

These dolphins are capable of diving to depths of 60 m (200 ft) and holding their breath for up to ten minutes. Atlantic spotted dolphins are also found in the shallower waters of the continental shelf and come in close to shore to feed near sandbanks or surf the bow waves of recreational boats. Their diet consists mainly of small fish, as well as bottom-dwelling invertebrates, squid, and octopus. They use echolocation for navigating and finding food, and pods often cooperate to hunt, encircling prey to

prevent them escaping. They can often be seen feeding alongside bottlenose, striped, and common dolphins, with which they regularly interact. Coastal pods may consist of as few as five to 15 dolphins, but the average is closer to 50. Occasionally, however, groups merge to travel in "super pods" of 200 or more.

Highly communicative, Atlantic spotted dolphins use whistles, squeaks, clicks, and buzzes as well as bubbles to "speak" to each other within a group, both at close and at long range. They also use pulsed barks or "screams" to warn off rival pods.

### Group activity
Pods are seen socializing in the clear shallow waters off the coast of the Bahamas. Through socializing, young males form strong, often lifelong, bonds with other males.

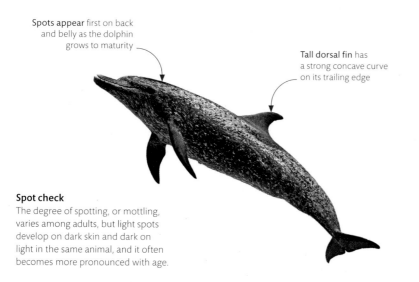

**Spots appear** first on back and belly as the dolphin grows to maturity

**Tall dorsal fin** has a strong concave curve on its trailing edge

**Spot check**
The degree of spotting, or mottling, varies among adults, but light spots develop on dark skin and dark on light in the same animal, and it often becomes more pronounced with age.

# the open oceans

**Many organisms** live in the challenging environment of the open oceans. Strong currents and a lack of shelter have led both predators and prey to evolve adaptations for greater speed or to camouflage themselves.

**Inflated float**
Some siphonophores – such as the Arctic species *Marrus orthocanna* – stay buoyant with the aid of a gas-filled sac. By adjusting gas levels in the sac, the animal can control its level in the water.

**Bright orange pneumatophore** (gas sac) is situated adjacent to the propulsive bells

# dividing labour

A group of animals called siphonophores swim or drift in the ocean like true jellyfish, yet they live in colonies like corals. Each colony is made up of individuals, called zooids, that are adapted to perform different tasks. This division of labour improves the overall efficiency of the colony. Some zooids attached to the colony's stem provide propulsion and function like the pulsating bells of a jellyfish; other zooids, with a specialized feeding role, have mouths and tentacles like the polyps of a coral or anemone.

**Mid-water colony**
Two swimming bells of a deep-ocean, cup-bearing siphonophore (*Sulculeolaria biloba*) are attached to a soft trailing stem that carries feeding polyps with tentacles for catching planktonic prey. One of the bells has an oil droplet to help maintain buoyancy.

## COLONIAL ORGANIZATION

The zooids of a siphonophore share nutrients harvested from planktonic prey through a common stem that is differentiated into two regions: one, the nectosome, with pulsating bells; and another, the siphosome, with polyps and tentacles. A gas-filled float, or pneumatophore, may also be present. In some siphonophores, such as the Portuguese man o' war, the swimming nectosome is absent and the colony floats on the ocean surface by its pneumatophore, drifting as it is blown by the wind.

Pneumatophore

Nectosome

Pulsating bell

Feeding polyp

Siphosome

Stem

Reproductive polyp

**GENERALIZED SIPHONOPHORE STRUCTURE**

## Diatoms

These are single-celled organisms whose cell walls are made from opaline silica. They are responsible for the production of much of the world's oxygen, and for removing carbon dioxide from the atmosphere. The silica in the cell walls make them heavier than other single-celled organisms, but adaptations enable them to remain at the water's surface for photosynthesis. They are also kept in suspension thanks to water turbulence.

**Curved cells** link to form long, spiralling chains 30–300 micrometres (millionths of a metre) long

**Body** is composed of two halves that fit together like a tiny box

**Flat, disc-shaped** cell that lives singly or in colonies

**GUINARDIA STRIATA**

**TRICERATIUM FAVUS**

**PLANKTONIELLA SOL**

## Dinoflagellates

These are organisms with two flagella (whiplike tails) that help them move through water in a corkscrew fashion. Most have bodies covered with complex outer cell walls. Some kinds can catch and absorb other plankton. When too many nutrients are available, large blooms may form, which leads to a phenomenon called "red tides", toxic to both marine life and humans.

**Bioluminescent species** with numerous chloroplasts; cell size can be 30–1,000 micrometres

**Cells** lack hard outer cell walls

**Tentacle** used to sense and capture prey

**PYROCYSTIS PSEUDONOCTILUCA**

**GYMNODINIUM**

**NOCTILUCA SCINTILLANS**

# phytoplankton

Organisms that drift with ocean currents and make their own food by photosynthesis are known as phytoplankton. They include microscopic algae and cyanobacteria. As phytoplankton are reliant on sunlight for photosynthesis, they are confined to the upper layers of the oceans. They are vital both to the global carbon cycle (they remove carbon dioxide and release oxygen) and to marine food webs, providing food for all forms of marine life from zooplankton through to whales.

**Fanned colony**

A brackish water diatom, *Licmophora flabellata* is named for the distinctive fanlike formation of the colony of cells. Adjacent cells are linked at the tips of the main axes, but by forming a "chain" they increase their overall surface area. A benthic species often attached to red and brown seaweeds, it is also found in coastal waters.

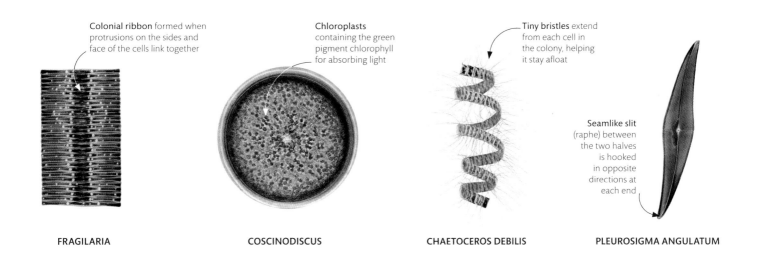

**Colonial ribbon** formed when protrusions on the sides and face of the cells link together

**Chloroplasts** containing the green pigment chlorophyll for absorbing light

**Tiny bristles** extend from each cell in the colony, helping it stay afloat

**Seamlike slit** (raphe) between the two halves is hooked in opposite directions at each end

**FRAGILARIA**

**COSCINODISCUS**

**CHAETOCEROS DEBILIS**

**PLEUROSIGMA ANGULATUM**

**Horns** formed by cell walls enable organism to spin as it moves in water

**Moon shape** gives rise to the scientific name of this widely distributed species

**Fans** are formed from many narrow, wedge-shaped cells

**CERATIUM HIRUNDINELLA**

**TRIPOS LUNULA**

**Individual cells** are no more than 0.25 mm long

Egg capsules suspended beneath the raft contain the developing larvae

**Floating nursery**

A floating egg mass of the violet snail (a species of *Janthina*) – here viewed from below – is carried by the drifting mother until the eggs hatch into swimming larvae. The larvae spend their early lives with other planktonic animals. As it matures, each larva then produces a thread of mucus that is tipped with air bubbles. This buoys the youngster upwards and brings it to the surface, where it lives out the rest of its days as a bubble-rafting adult.

Each bubble takes about 10 seconds to make, as the snail envelopes a pocket of air with its muscular foot

The floating raft is composed of dozens of air-filled bubbles, each held in place by a thin coating of dried mucus

**Buoyant snail**
A violet snail clings tightly to its bubble-raft with its muscular foot, while the rest of the body hangs beneath. The snail's shell is especially thin and lightweight, which helps to keep it afloat.

The shell is helical in shape, just like the shells of many more typical snail species

# staying afloat

The ocean surface, where Sun-fuelled plankton is at its most abundant, can be a profitable place to feed. The community of surface animals, called the pleuston, has some unlikely floating members. Instead of crawling over the seabed, violet snails drift in the open sea and prey on buoyant hydrozoans. They use their foot to mould mucus-coated bubbles of air that dry to form an inflated raft. This structure, though delicate, can support an adult snail, even when carrying a clutch of eggs.

**PREDATORS AND PREY**
Many colonial hydrozoans, which are relatives of jellyfish and anemones, drift among the pleuston supported by gas-filled bladders. Among them are the Portuguese man o'war (*Physalia physalis*) and the by-the-wind sailor (*Velella velella*), which use stinging tentacles to ensnare planktonic prey. Other members of the pleuston – including violet snails – are predators of both.

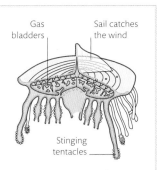

Gas bladders

Sail catches the wind

Stinging tentacles

**CROSS SECTION OF BY-THE-WIND-SAILOR**

# sustaining life in the dark

Beneath volcanic parts of the dark ocean floor, sea water seeping downwards meets hot, molten rock, causing the water to belch back out of fissures, called hydrothermal vents, as boiling streams of concentrated brine. Bacteria use chemicals in the jets of vent water: the only deepsea organisms to do so. This process supports entire communities of animals around the vents, including swarms of shrimps and giant worms. Vent animals are independent of the light-powered food chains at the ocean surface and on land, which start with algae and plants.

**A pair of pinkish "eyespots"** under the carapace detects infrared radiation from vents and guides the shrimp to places where it can find food

**The thin, transparent carapace** enables more infrared radiation from hydrothermal vents to reach the elongated "eyespots" beneath

## HYDROTHERMAL VENTS

As hot, chemically rich brine from vents meets the cold, deep ocean, minerals such as iron sulphide solidify into chimneys and clouds of black "smoke". Bacteria on the chimney walls process hydrogen sulphide in the brine and oxygen from the surrounding water to release energy. They use the energy to make sugar and other nutrients from carbon dioxide and water, in a process called chemoautosynthesis.

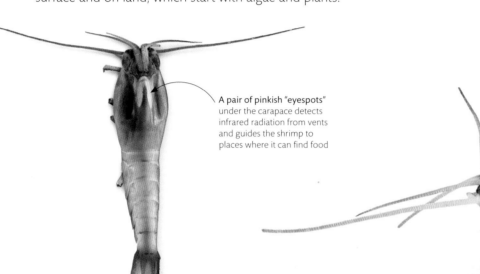

Sea water seeps down through the seafloor

Geothermal activity heats the sea water, which mixes with sulphides and gases from the surrounding rock

Hydrogen sulphide reacts in nearby bacteria to release energy

Iron sulphide forms chimneys or escapes as "smoke"

Hot water shoots up through vents in the seafloor

**HOW A HYDROTHERMAL VENT SUPPORTS A FOOD CHAIN**

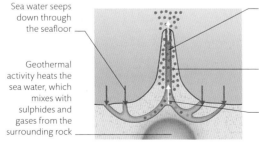

**Ocean-floor shrimp**
The mid-Atlantic ridge shrimp (*Rimicaris exoculata*) grazes on food-making bacteria and organic detritus in the warm water around hydrothermal vents, often more than 3 km (1.8 miles) deep. In places, this species gathers in hordes of up to 2,500 individuals per square metre.

**Red gills** absorb the gases – hydrogen sulphide, oxygen, and carbon dioxide – needed by bacteria living in the tube worm's body

**Symbiosis**
Some animals living near hydrothermal vents, such as giant tube worms (*Riftia pachyptila*), lack digestive systems for consuming food. Instead, they rely on food-making bacteria living inside their bodies to generate all of their nourishment.

**Thoracic chamber** contains the reddish-brown gills, which are colonized by food-making bacteria after each moult; the bacteria probably supplement food that the shrimp obtains by grazing

**At the junction**
A diver descends into the rift between
the American and Eurasian tectonic plates.
Iceland's Silfra Canyon, which is filled with
fresh water, is one of the few parts of the
mid-ocean ridge that has been explored.

# new ocean floor

The global mid-ocean ridge system is by far the longest mountain
range on Earth, extending for over 80,000 km (50,000 miles)
under every ocean worldwide. The ridge is broad and rugged,
rising up to 3 km (2 miles) above the abyssal plain (the seafloor
between the continental margins and mid-ocean ridge). It consists
of a continuous string of active volcanoes, and follows the line
of divergent plate boundaries, where tectonic plates move apart.
Earthquakes occur frequently along its length. New ocean floor
is formed along the ridge from the upwelling of volcanic material.
The vast area of the mid-ocean ridge remains almost completely
unexplored. Rarely, the crest of the ridge rises above the sea – for
example, in Iceland, where volcanic activity is especially intense.

## THE MID-OCEAN RIDGE

New oceanic crust is formed along the mid-ocean ridge where molten lava
(magma) extrudes onto the seafloor, over 2 km (1.2 miles) beneath the ocean
surface. Molten magma continually oozes out through fractures in the crust
and cools rapidly to form mounds known as pillow lava. Vertical sheets of dark
basalt rock (dikes) harden in the fractures and slowly force apart the two sides
of newly formed ocean floor.

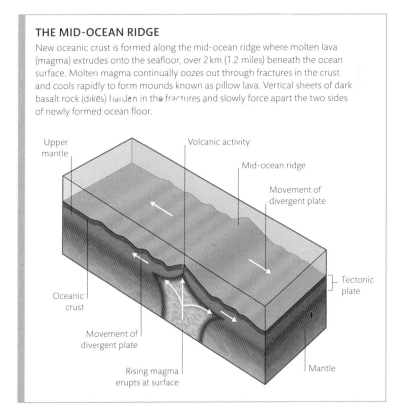

Upper mantle

Volcanic activity

Mid-ocean ridge

Movement of
divergent plate

Oceanic
crust

Tectonic
plate

Movement of
divergent plate

Mantle

Rising magma
erupts at surface

Each armoured plate that makes up the body's exoskeleton can resist the high pressure at great ocean depths

Head capsule is part of the exoskeleton that acts like a helmet and protects the brain

Antennae, each growing up to 20 cm (8 in) long, contain tactile and chemical sensors that help the isopod to navigate in the dim light on the ocean bottom

A robust, curved claw is carried on each of the 14 walking legs to help grip the ocean floor as well as large food items, including carrion

**Armoured giant**

The most familiar of the isopods are their small terrestrial representatives: woodlice. But some supergiant species – such as the western Atlantic giant isopod (*Bathynomus giganteus*), from below 800 m (2,600 ft) – can grow to 50 cm (20 in) long. Like a seabed armadillo, it has tough segmented armour that protects it from bigger predators.

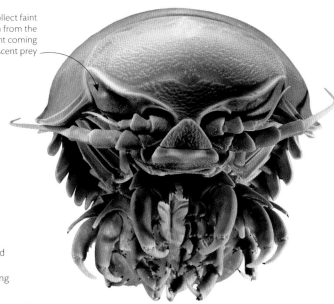

**Compound eyes** collect faint sunlight reaching down from the surface, as well light coming from bioluminescent prey

**Seafloor opportunist**
Food is scarce at the bottom of the ocean, so the scavenging giant isopod may occasionally turn predatory by using its jointed legs to grab a creeping sea cucumber or slow-moving fish.

# deepwater giants

The largest animals on the planet live in the ocean. The champion of them all, the blue whale, weighs as much as 40 African elephants – the biggest living animals on land. Bodies can grow more when supported by water's buoyancy (see p.212), and although growth is slow in the cold of the greatest ocean depths, some deepwater animals live long enough to become giants of their kind. Here, there are cousins of woodlice that grow as large as miniature dogs, and squid that can achieve the length of a bus.

### OVERSIZE SQUID

Reaching up to 10 m (33 ft) in length, the colossal squid (*Mesonychoteuthis hamiltoni*) is the world's largest invertebrate. Giant squid (*Architeuthis*) have longer tentacles but smaller bulk. Despite their size, all these predatory behemoths – typically found at depths of more than 500 m (1,600 ft) – are rarely seen alive. Much of what is known about them comes from specimens washed ashore.

**STRANDED GIANT SQUID, NEWFOUNDLAND, 1883**

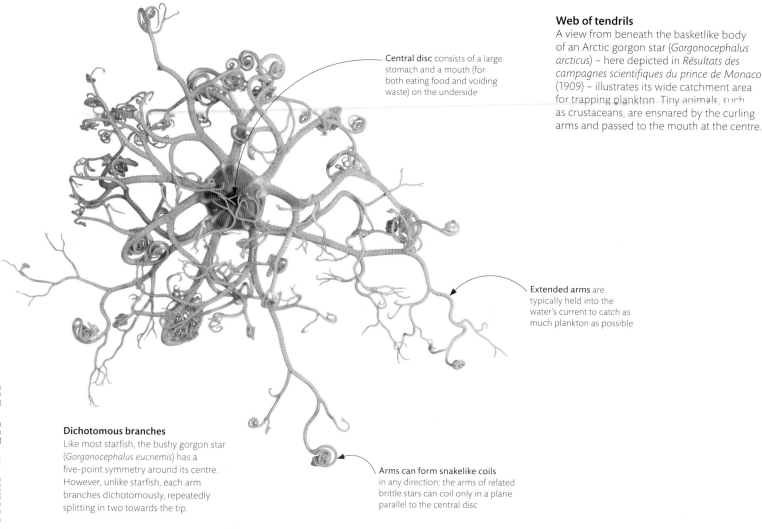

Central disc consists of a large stomach and a mouth (for both eating food and voiding waste) on the underside

**Web of tendrils**
A view from beneath the basketlike body of an Arctic gorgon star (*Gorgonocephalus arcticus*) – here depicted in *Résultats des campagnes scientifiques du prince de Monaco* (1909) – illustrates its wide catchment area for trapping plankton. Tiny animals, such as crustaceans, are ensnared by the curling arms and passed to the mouth at the centre.

Extended arms are typically held into the water's current to catch as much plankton as possible

**Dichotomous branches**
Like most starfish, the bushy gorgon star (*Gorgonocephalus eucnemis*) has a five-point symmetry around its centre. However, unlike starfish, each arm branches dichotomously, repeatedly splitting in two towards the tip.

Arms can form snakelike coils in any direction; the arms of related brittle stars can coil only in a plane parallel to the central disc

Arms coiled tightly around the central disc during daylight are less vulnerable to predators

# branching arms

The radial arrangement of arms found in starfish and their relatives allows these animals to creep over the seabed, with movement largely restricted to the horizontal plane. But gorgon stars, cousins of brittle stars (see pp.218–19), can also reach upwards into the water column. Their arms branch into a writhing mass of tendrils that is reminiscent of the eponymous Greek monster. The intertwined tendrils are an effective trap for catching small planktonic animals suspended in the water – a source of food beyond the reach of starfish and brittle stars.

**Resting by day**
Gorgon stars, such as this spiny gorgon star (*Astrocaneum spinosum*), keep their arms retracted during the day. They extend their arms at dusk and feed throughout the night.

## Hydrozoa

This is a very large group of animals that are seen as distinct from true jellyfish. As well as solitary organisms with a bell-shaped medusa form, the group includes some colonial forms known as siphonophores, which consist of polyps and sometimes attached medusae (missing from the Portuguese man o' war). Other hydrozoans may have medusa or polyp forms absent from their life cycle; the eight-strand jelly has a polyp stage, but the golf tee jelly does not.

**Mouth** with eight folded lips leads to short, octagonal stomach

**EIGHT-STRAND JELLY**
*Melicertum octocostatum*

**Circular membrane** at bell base reduces aperture increasing force of water jet for propulsion

**GOLF TEE JELLY**
*Aeginopsis laurentii*

**Gas-filled float** (pneumatophore) keeps colony afloat

**Long, beadlike tentacles** capture and immobilise prey; each bead contains stinging cells (nematocysts)

**PORTUGUESE MAN O' WAR**
*Physalia physalis*

## Cubozoa

Also known as box jellies or sea wasps, cubozoa are named for their box-shaped medusa, and are the most venomous cnidarians. They have muscular pads, or pedalia, at each corner of the bell canopy, with one or more tentacles attached to each one. The attached polyp stage of the life cycle metamorphoses into a medusa, rather than budding to release multiple medusae as in true jellyfish.

**Bell** is only 2 cm (¾ in) in diameter, but the stinging capsules are the most deadly

**IRUKANDJI JELLY**
*Carukia barnesi*

**Sense organs** (rhopalia) are clustered at the base of each tentacle

**WINGED BOX JELLY**
*Alatina alata*

**Tentacles** are covered with millions of nematocysts that release highly venomous darts

**COMMON BOX JELLY**
*Chironex fleckeri*

## Staurozoa

This group, known as stalked jellyfish, includes animals whose life cycle is more similar to that of a coral or anemone than other jellyfish. They have a stalked, trumpet-shaped body and do not alternate between polyp and free-swimming medusa. They attach themselves to a substrate such as coral, from which they hang for the rest of their lives.

**Each arm**, or branch, is equipped with 100–140 tentacles

**HORNED STALKED JELLYFISH**
*Lucernaria quadricornis*

**Arms** are connected near the tips by a thin membrane

**KALEIDOSCOPE STALKED JELLYFISH**
*Haliclystus auricula*

**Funnel-shaped** "head" (calyx) is the same length as its stalk (peduncle)

**THICK-RIMMED STALKED JELLYFISH**
*Haliclystus salpinx*

## Scyphozoa

Also known as true jellyfish, this is a group of about 200 free-swimming marine species. Polyps of these animals may be less obvious, but are often long-lived. They are most conspicuous in their medusa stage – that is, with an umbrella-shaped body – but this may not form the longest part of their life cycle. They move by contracting and relaxing the muscles of their umbrella.

**Colourful jelly** that emits light when disturbed

**Four lobes** (oral arms) hang down from mouth

**MAUVE STINGER OR NIGHT LIGHT JELLY**
*Pelagia noctiluca*

**Short tentacles** form a fringe around bell margin

**BROWNBANDED MOON JELLY**
*Aurelia limbata*

**Bell** can be 60 cm (2 ft) across, with 3-m- (10-ft-) long tentacles

**NORTHERN SEA NETTLE JELLY**
*Chrysaora melanaster*

**Saucer-shaped bell** is thick in the centre and thins out towards the edge

**Mane of hairlike** stinging tentacles each up to 36 m (120 ft) long entangle the prey

## Giant of the oceans

The largest true jellyfish species in the world, the lion's mane jelly (*Cyanea capillata*) is a continual swimmer that can cover vast distances in strong marine currents. Most are found in the Arctic, north Atlantic and north Pacific oceans, where they live on a diet of zooplankton, fish, Moon jellyfish, and shrimp.

# jellyfish and hydrozoans

These mostly free-swimming, jellylike predatory animals belong to the group of invertebrates known as cnidarians. Most jellyfish and hydrozoans alternate between polyp and medusa (bell- or umbrella-shaped) forms. Like other cnidarians, these animals have radial symmetry and lack a heart and a brain, but use an internal cavity to digest prey, which they catch using tentacles armed with stinging cells.

**Ram filter-feeding**
A large whale shark swimming at 1 m (3 ft) per second can filter around 1 million litres (264,000 gallons) of sea water per hour.

**Mouth** contains vestigial teeth and spongy filter pads with a mesh size of about 1 mm

# filter-feeding sharks

Filter-feeding is an efficient feeding system that enables some of the largest animals in the ocean, including baleen whales and some sharks (basking, megamouth, and whale sharks), to feed on some of the smallest. All three shark species feed mainly by passive ram filtration, which involves swimming slowly, with an open mouth, through swarms of plankton and small squid and fish. The food is then filtered out of the water. The whale shark (*Rhincodon typus*), which is by far the largest living fish at 9–15 m (29–49 ft) long, can also generate suction to feed selectively on larger items.

**Migrating to feed**
Whale sharks travel huge distances between their breeding areas and feeding waters. Here the shark is accompanied by a remora. These fish attach to sharks by suction using their dorsal fin and feed off the sharks' faeces and ectoparasites.

## CROSS-FLOW FILTRATION

The filter-feeding system of the whale shark is known as cross-flow filtration. The feeding current flowing from the mouth passes parallel to the filter pads, where water is drawn out sideways over the gills. Food particles become concentrated into a swallowable mass at the back of the throat. This system is less prone to clogging than one in which the flow hits the filter head on, although sharks do sometimes "cough" to backwash material from their filters.

**WHALE SHARK'S FILTER-FEEDING SYSTEM**

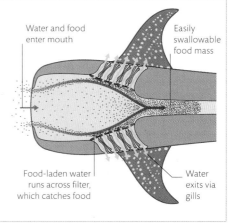

Water and food enter mouth

Easily swallowable food mass

Food-laden water runs across filter, which catches food

Water exits via gills

# ocean currents

Wind is the principal force that drives currents in the ocean surface. It works in conjunction with pressure gradients caused by gigantic mounds of sea water built up in a series of wind driven loops, known as gyres, and by the effect of Earth's rotation – the Coriolis effect. These currents are hugely powerful. For example, the flow of the North Atlantic Gulf Stream alone is equivalent to the total discharge of the world's top 20 rivers added together. Equally important is a vast network of slow-moving, deep currents, involving 90 per cent of ocean water. These currents circulate energy, nutrients, salt, and sediments around the world. Surface and deep-sea currents act together to moderate Earth's climate.

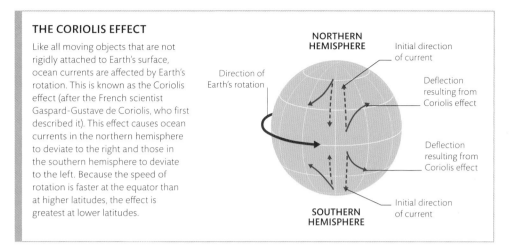

### THE CORIOLIS EFFECT

Like all moving objects that are not rigidly attached to Earth's surface, ocean currents are affected by Earth's rotation. This is known as the Coriolis effect (after the French scientist Gaspard-Gustave de Coriolis, who first described it). This effect causes ocean currents in the northern hemisphere to deviate to the right and those in the southern hemisphere to deviate to the left. Because the speed of rotation is faster at the equator than at higher latitudes, the effect is greatest at lower latitudes.

NORTHERN HEMISPHERE

Direction of Earth's rotation

Initial direction of current

Deflection resulting from Coriolis effect

Deflection resulting from Coriolis effect

Initial direction of current

SOUTHERN HEMISPHERE

**Perpetual ocean**
Using data from numerous sources, NASA compiled a visualization of Earth's ocean currents. This segment of the Caribbean and western North Atlantic shows the Gulf Stream, which flows northeastwards from Florida.

# white shark

An oceanic apex predator, also known as the great white, the white shark (*Carcharodon carcharias*) can be more than 6 m (20 ft) long and weighs several tons. White sharks were once feared by humans, yet attacks are rare and humans pose the greatest threat to this officially vulnerable species.

Although increasingly rare, white sharks have one of the widest geographic ranges of any marine animal and can be found in all cold, temperate, and tropical waters. This slow-growing shark takes up to 16 years to reach maturity, and females only give birth every two to three years. Humans' hunting for sharks' jaws, fins, and teeth, and attempts to reduce numbers near beaches, have exacted a heavy toll on their population. However, the white shark possesses remarkable characteristics that make it a formidable hunter, increasing its chances of survival.

It is partially endothermic, or warm-blooded, so able to maintain its brain, swimming muscles, and stomach at a temperature higher than its ocean habitat. This not only keeps the animal highly active but also helps power astonishing bursts of speed of up to 60 kph (37 mph) when hunting. The white shark also has a streamlined body and excellent colour vision, as well as the largest scent-processing organs of any shark. Highly sensitive electroreceptors on its head can detect even the slightest of prey-generated electrical fields. In spite of this, the white shark often relies on ambush tactics when hunting. It typically swims slowly beneath potential prey, which are outlined against the ocean's surface. However, the prey cannot see their predator because the shark's dark upper body blends in with the murky depths. When attacking from below, a shark makes a near-vertical strike, often leaping 8 m (26 ft) clear of the water.

### Tools of the trade

This shark has narrow bottom teeth to hold its kill and large upper ones to tear flesh. It favours high-calorie prey such as seals or dolphins, but it eats squid, turtles, fish, and other sharks, too.

### REPLACEMENT TEETH

Adult white sharks may have up to 300 retractable teeth, arranged in up to seven rows, at any one time. Attached to the jaws only by soft tissues, not sockets, the teeth are frequently lost due to pressure exerted when the shark bites its prey. When this happens, a tooth from the row behind moves forwards to replace it. A white shark can go through 20,000 or more teeth in its lifetime.

Front row

Second row

Soft tissue connecting tooth to jaw

Lower jaw

**LOWER JAW WITH ROWS OF TEETH**

**Dorsal fin**, working in synchrony with the symmetrically arranged anal fin, acts as a hydrofoil: as it flaps, it provides some lift as well as a little propulsion

**Clavus** – a thick, wide frill around the rear of the fish – probably forms from extensions of the dorsal and anal fins that replace the tail lost during development

## Tailless wonder

The truncated shape of a giant ocean sunfish (*Mola mola*) arises because the vertebrae and fin of the tail are missing – replaced instead with a leathery flap, or clavus. This works like a rudder, while the vertical dorsal and anal fins flap from side to side to move the fish very slowly forwards.

**Cleaner wrasse** feeds on parasites infesting the skin; the ocean sunfish itself lacks the agility to dislodge parasites easily

**Spiny beginnings**
The protective spiny skin of
ocean sunfish fry is reminiscent
of pufferfish – their close relatives.
Ocean sunfish fry hatch from eggs
with a working tail, which shrinks
as they develop.

**Wide pectoral fins**, which
stabilize the body, are
retained into adulthood

# slow giant

Plankton are organisms that drift in water because they
lack the strength to swim against currents. Most are
tiny, but some much larger animals have such weak
powers of propulsion that they may sometimes move
like plankton. They include the heaviest bony fish:
ocean sunfish. Weighing as much as a rhinoceros, they
lack any trace of the sweeping tail most other fish use to
generate thrust.

**Thick, scaleless skin** overlies a
layer of low-density gelatinous
tissue that maintains buoyancy,
even though the fish lacks a gas-
filled swim bladder

## FLOATING EGGS

Pelagic (open-ocean) fish are adapted to live at different depths, but
most produce buoyant eggs that rise to the surface. Here they hatch
into fry that spend their infancy feeding on the microscopic, Sun-
fuelled plankton. Producing many eggs increases the chance that
some will survive. A single ocean sunfish may release 300 million
eggs – a record among vertebrates.

Oil in eggs gives
buoyancy

Ocean
sunfish larva

Bristlemouth
larva

Ocean sunfish
lives near the
surface

Depth (m)

1,000

Bristlemouth
lives in the
deep ocean

2,000

As larvae mature,
deepwater species
swim down to
adult depths

3,000

4,000

**EGG, LARVAL, AND ADULT DEPTHS OF PELAGIC FISH**

**Carta Marina (1539)**
The largest and most accurate map of Scandinavia
of its time was produced by exiled Catholic priest
Olaus Magnus, whose mission was to share the
history, culture, and natural wonders of the north
with the rest of Europe. It was printed from nine
carved wooden blocks and shows Nordic lands
and seas teeming with life, including the monsters
described to Magnus by fishermen and sailors.

**New Map of the Whole World (1648)**
One of the great works of the Dutch Golden Age of cartography, Joan Blaeu's breakthrough map *Nova totius terrarum orbis tabula* is part of a huge atlas showing the discovered world fairly accurately, with northwest America and Australasia indicated in incomplete outlines.

*the ocean in art*

# mapping the seas

Since modern humans first carved designs into rocks up to 40,000 years ago, people have used their cognitive spatial skills to map themselves in relation to their environment. By the 15th century's Age of Exploration, an increasing number of maps from around the world emerged as Europeans recorded territories they travelled to in pursuit of trade and colonization.

Primitive maps carved in stone or wood, or created in 3D reliefs from sand and clay, were superseded over time by woodblock printing, then copperplate printing in the 16th century and lithography in the 19th century. Medieval map-makers based the proportions of countries on their relative wealth and power. However, this changed in the 14th century, when the works of the Greek geographer and astronomer Ptolemy (*c.*100–*c.*170 CE) were rediscovered and translated into Latin. His calculations of the size and proportions of Earth changed Western cartography forever, although they later proved to be incorrect.

In 1539, Swedish priest Olaus Magnus produced a map of Scandinavia (see left) – an attempt to correct Ptolemy's vague

rendering of northern Europe produced in distant Mediterranean climes. The geography is immediately recognizable, and even the whorls in the ocean may represent known currents and fronts, but what enchants is the detail. The everyday lives of the Scandinavians – driving sleighs, milking reindeer, and spearing seals –are played out in a natural world teeming with real and fantasy beasts. None more so than in the seas, where a dragon wrestles with a giant crayfish, ships anchor mistakenly on tusked whales, and a red sea serpent is described in the map's key as "a worm 200 feet long wrapping itself around a big ship and destroying it".

A century later, Dutch cartographer Joan Blaeu's remarkable map of the world (see above) showcases the great copperplate engraving skills of the Dutch Golden Age. It also reflects the nation's prowess in seafaring and the artist's considerable scientific knowledge. The work features the five known planets in personified form with the Sun at their centre, endorsing Copernicus's controversial heliocentric theory that Earth orbits the Sun.

> **“** Inside this broad expanse of fluid Ocean ... a conglomeration of monsters may be found. **”**

OLAUS MAGNUS, *A DESCRIPTION OF THE NORTHERN PEOPLES (VOLUME III)*, 1555

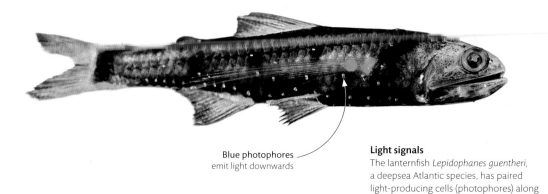

**Blue photophores**
emit light downwards

**Light signals**
The lanternfish *Lepidophanes guentheri*, a deepsea Atlantic species, has paired light-producing cells (photophores) along its sides and on its head. These lights are used in communication and courtship.

# lights in the dark

Bioluminescence, a process in which living organisms produce their own light (see pp.192–93), is particularly common in the unlit waters of the deep oceans. Deepwater fish may use bioluminescence for several purposes. For predatory fish, such as some anglerfish, attracting prey with a luminous lure on a chin barbel or suspended on a structure like a fishing rod can be an energy-efficient way to hunt. Other species use light to disguise their outline (see below), confuse predators, or attract a mate.

### LIGHT AS CAMOUFLAGE

Some fish, such as the hatchetfish (shown here), have light-emitting photophores on their underside. These help to disguise the fish's silhouette when viewed from below – a camouflage method known as counter-illumination. The fish can control the intensity of the light that it produces to match that of sunlight filtering down from above, making the fish less visible to predators.

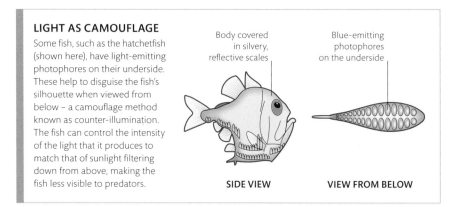

Body covered in silvery, reflective scales

Blue-emitting photophores on the underside

**SIDE VIEW**

**VIEW FROM BELOW**

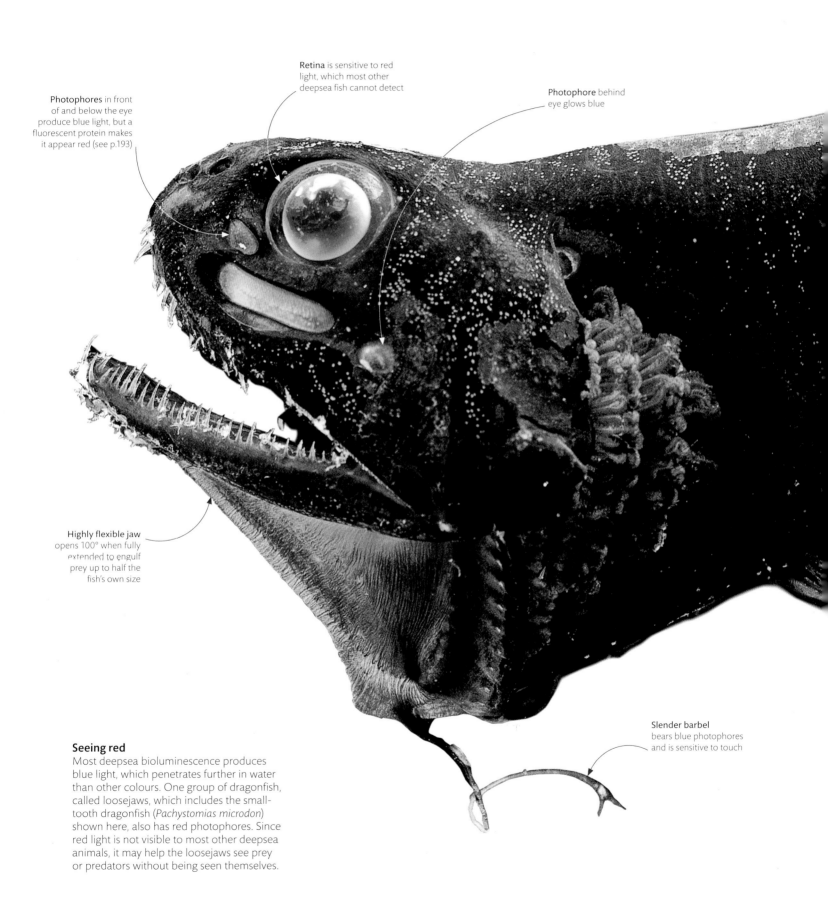

**Retina** is sensitive to red light, which most other deepsea fish cannot detect

**Photophores** in front of and below the eye produce blue light, but a fluorescent protein makes it appear red (see p.193)

**Photophore** behind eye glows blue

**Highly flexible jaw** opens 100° when fully extended to engulf prey up to half the fish's own size

**Slender barbel** bears blue photophores and is sensitive to touch

### Seeing red

Most deepsea bioluminescence produces blue light, which penetrates further in water than other colours. One group of dragonfish, called loosejaws, which includes the small-tooth dragonfish (*Pachystomias microdon*) shown here, also has red photophores. Since red light is not visible to most other deepsea animals, it may help the loosejaws see prey or predators without being seen themselves.

Tail sweeps rapidly from side to side, helping to maximize thrust and minimize drag

Huge, sail-like dorsal fin, which extends almost the length of the body, is raised for hunting

Trunk is streamlined and muscular, helping sailfish to reach top speeds when swimming

### Cornering at speed

When hunting, the sailfish (*Istiophorus platypterus*) raises its large dorsal fin, which aids manoeuvrability and provides stability so the fish can improve its aim. It hunts by probing schools of smaller fish with its long, sharp bill, then swiping suddenly sideways, often stunning or maiming several prey at a time.

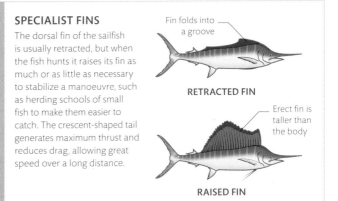

**SPECIALIST FINS**

The dorsal fin of the sailfish is usually retracted, but when the fish hunts it raises its fin as much or as little as necessary to stabilize a manoeuvre, such as herding schools of small fish to make them easier to catch. The crescent-shaped tail generates maximum thrust and reduces drag, allowing great speed over a long distance.

Fin folds into a groove

**RETRACTED FIN**

Erect fin is taller than the body

**RAISED FIN**

Elongated, bony bill is an extension of the snout

# built for speed

Billfish, such as sailfish, marlin, and swordfish, benefit from two key weapons: a long, serrated bill and the ability to swim at great speed. Their bodies are highly streamlined and packed with muscle for both rapid acceleration and endurance. The sailfish is one of the fastest fish, with a top speed of about 35 kph (22 mph) per hour in short bursts.

Long pectoral fins can help to generate lift when extended

Torpedo-shaped keel directs flow of water over the tail

**Tail keels**
Small, horny ridges (keels) on the sailfish's tailstock make swimming more streamlined and provide stability. The tail beats side to side up to eight times a second.

# wandering albatross

One of the world's largest flying birds, the wandering albatross (*Diomedea exulans*) can be 1.3 m (4 ft 3 in) long and weigh as much as 12 kg (26 lb). This species is found mainly in the circumpolar region of the Southern, or Antarctic, ocean and has a lifespan of 50 years or more.

**Elaborate courtship**
The albatross's display includes wing-spreading, head-waving, bill-rapping, and a distinctive braying call.

The wandering albatross has the longest wingspan – up to 3.5 m (11 ft 6 in) – of any bird. It is a consummate glider that flies vast distances in search of squid and other cephalopods. A bird can cover up to 10,000 km (6,200 miles) in 10 to 20 days. It often soars for several hours without flapping its wings by gliding on thermals and allowing winds to push it along. An albatross spends most of its life at sea; during its first six years, it may never touch land, although it might rest on the ocean occasionally to digest a heavy meal or when there is not enough wind to fly. It can drink sea water, as it secretes excess salt via a gland above the nasal passage.

Sexually mature at around the age of 11, albatrosses mate for life, but they breed only once every two years. The female lays one egg in a nest of mud and grass, usually on subantarctic islands, and the parents take it in turns to sit on the nest.

Unfortunately, human fishing equipment kills thousands of these birds each year, placing the survival of all albatross species in jeopardy.

**Soaring across the waves**
Albatrosses can cover 1,000 km (620 miles) per day, and they have been recorded reaching speeds of 108 kph (67 mph), yet they consume only slightly more energy in flight than when sitting on their nests.

**Beneficial feeder**
The recovery of great whale populations, including
that of the endangered blue whale (here off the
coast of Mexico), may play a key role in averting
climate change. Their faeces are rich in iron, which
is released from krill during digestion. This boosts
the growth of phytoplankton, which in turn absorb
and store vast quantities of carbon.

# bulk feeding

There are two main types of whale: toothed whales and
baleen whales, which are filter-feeders and have baleen
plates instead of teeth. The blue whale (*Balaenoptera
musculus*), the largest animal ever to have lived on Earth,
belongs to the family of baleen whales known as
rorquals, distinguished by the grooves or pleats on their
throat, which expand to accommodate vast volumes of
food-laden sea water. When swallowing a dense swarm
of krill, a blue whale can engulf 200,000 litres (53,000
gallons) of water and almost half a million food calories.

## HOW BALEEN WORKS

Baleen whales have hundreds of baleen plates hanging from their
upper jaw. These plates are fringed with stiff hairs, which form a
"curtain" that prevents food, mainly shrimplike krill, from escaping
with sea water as it is squeezed out of the closing mouth.

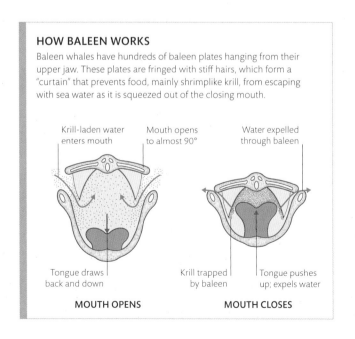

Krill-laden water
enters mouth

Mouth opens
to almost 90°

Water expelled
through baleen

Tongue draws
back and down

Krill trapped
by baleen

Tongue pushes
up; expels water

**MOUTH OPENS**          **MOUTH CLOSES**

**Ancient activity**
The Tao-Rusyr caldera off Onekoten Island, Kamchatka, was formed by volcanic activity that occurred in the northwestern sector of the Pacific Ring of Fire over 9,000 years ago.

# destruction of the oceans

Earth's outer layers are divided into huge, mobile fragments called plates. The zones where plates collide, called subduction zones, are the world's largest recycling plants. Here old rocks in Earth's outermost layer, or crust, are drawn back down into the interior. Newer seafloor is produced at the same rate elsewhere, at mid-ocean ridges (see pp.264–65). Subduction zones encircle the Pacific Ocean, consuming the ocean floor at rates of up to 15 cm (6 in) per year. These are regions of great heat and pressure, and produce deep-seated earthquakes and volcanic eruptions in a band called the Ring of Fire. When they occur under the ocean, earthquakes and landslides can trigger devastating tsunamis (see pp.76–77).

## COLLIDING PLATES

Where two tectonic plates collide, such as an oceanic plate and a continental plate, the pressure forces the older one, which has a cooler, denser crust, beneath the younger plate. A deep ocean trench is formed where seafloor sediments are dragged downwards. As the plates grind past one another, enormous frictional strain is generated. The energy is released by periodic violent earthquakes, while the intense heat created begins to melt the rocks, and this molten magma pierces upwards to form a line of volcanic sea mounts, known as an island arc.

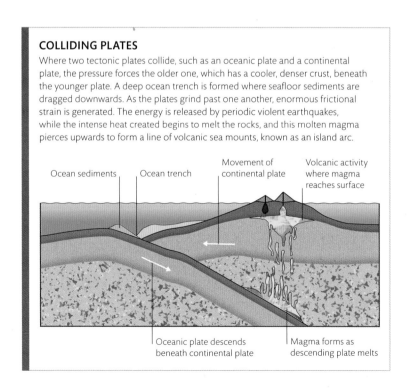

Ocean sediments    Ocean trench    Movement of continental plate    Volcanic activity where magma reaches surface

Oceanic plate descends beneath continental plate    Magma forms as descending plate melts

# orca

One of Earth's largest living predators, the orca (*Orcinus orca*) is also one of the fastest marine mammals, reaching speeds of about 55 kph (34 mph). Often called the "killer whale", at up to 10 m (33 ft) long it is actually the largest member of the dolphin family – and one of the most intelligent.

Found in every ocean, orcas have the widest distribution of any sea mammal on the planet. Striking black-and-white markings make orcas easy to spot, particularly as adult males bear the largest dorsal fin of any cetacean, with some reaching up to 1.8 m (6 ft) high.

The orcas' success is due in part to their range of sophisticated hunting techniques, which are taught to juveniles by adults. These include coordinated attacks on large whales, herding fish before stunning them with powerful tail strikes, and "wave washing" seals off ice floes. Orcas are the only cetaceans that regularly eat other marine mammals – including other dolphins.

Sea lions, octopuses, sea otters, rays, turtles, squid, and penguins are part of the orca's diet, but what individuals eat depends largely on what population type they belong to: "resident" orcas, which form the largest pods (or groups), mostly hunt fish, squid, and octopuses; "transient" orcas feed almost exclusively on marine mammals; while "offshore" orcas mainly hunt fish, especially sharks.

Pods range from just a few animals to groups of 50 or more and usually consist of a mature female and her female offspring and relations. Orcas can mate year round, but most matings occur during late spring and summer. Following a gestation period of around 17 months, the female gives birth to a single calf, which is usually born tail-first.

**Keeping in touch**
Pod members communicate with each other via a sophisticated series of high-frequency clicks and whistles, as well as "pulsed calls", which, to human ears, sound like screams.

**Wave-washing**
Working together, several orcas swim underneath an ice floe to create a large wave that will knock the seal off into the water.

# the polar oceans

**Low temperatures** in the polar oceans slow the movement and growth of plants and animals. Yet the water is rich in oxygen and nutrients, enabling marine life to thrive in this extreme environment.

# sea butterfly

A tiny zooplanktonic sea snail sporting twin translucent winglike appendages, the sea butterfly (*Limacina helicina*) is critical to the ecology of the Arctic seas. It feeds on zooplankton and phytoplankton, which it traps by casting a net of mucus.

Sea butterflies are pteropods, or free-swimming marine gastropods, that propel themselves with elegant parapodia (or "wings") adapted from a modified foot. Their 1–14 mm ($^1/_{25}$–$^1/_2$ in) thick shells are formed from aragonite, a type of calcium carbonate absorbed from sea water.

At the bottom of the Arctic food chain, sea butterflies create huge swarms in the upper reaches of open Arctic waters, where they consume plankton – even other sea butterflies. They, in turn, are eaten by whales and seals, which can be prey for polar bears.

### Weighty armour
Sea butterflies are normally protected by their dense aragonite shells. However, their shells are heavy, and the animals sink rapidly if not actively swimming or hanging from their mucus feeding-webs.

These minute animals have a short, one-year life cycle and, when they die, their shells fall to the seabed. This creates an effective carbon "sink" that keeps tons of carbon dioxide ($CO_2$) out of the atmosphere and helps to slow the rate of global warming. However, oceans absorb more than 25 per cent of the rising $CO_2$ emissions, which raises acidity levels (lowers their pH). Ocean acidification hinders the sea butterflies' ability to build their protective armour by reducing the speed at which their shells grow and the availability of aragonite. While a damaged or non-existent shell does not necessarily kill sea butterflies, it does leave them vulnerable to predators and disease. The size and health of the sea butterfly population is therefore a key indicator of how much ocean ecosystems are being affected by global warming.

### FLYING THROUGH WATER

Sea butterflies swim by moving their parapodia, or "wings", in the same figure-of-eight pattern seen with the wings of flying insects. Both animals generate lift on up- and downstrokes by peeling their wings apart at the start of each downstroke, and rotating each wing slightly during the upstroke.

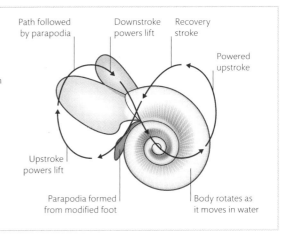

Path followed by parapodia

Downstroke powers lift

Recovery stroke

Powered upstroke

Upstroke powers lift

Parapodia formed from modified foot

Body rotates as it moves in water

**"WING" MOVEMENT MADE BY A SEA BUTTERFLY**

**Intricate patterns**
This aerial view of sea-ice off the
Icelandic coast shows small segments of
sea-ice (right) formed offshore combining
to produce larger expanses of ice (left) in
the shallower waters near the coast.

# sea-ice

The water in the ice-cold polar oceans freezes and thaws in an
annual cycle. During the winter months of 24-hour darkness,
temperatures can drop to to -30°C (-22°F) so that the sea freezes
over to form sea-ice. When water freezes, it is less dense than it is
as a liquid. Consequently, the ice that is formed floats. Each year,
this sea-ice covers a remarkable 40 million square km (15 million
square miles) of frozen ocean, which is equivalent to the area of
Russia, China, and the US combined. In summer, 60–80 per cent
melts back into the ocean, with no net effect on sea level. Any ice
that remains frozen through the summer grows thicker year-on-
year and drifts in the ocean currents, but the area of permanent
sea-ice is fast diminishing as global climate warms.

## HOW THE SEA FREEZES

Freezing winds across the ocean surface cause ice crystals to grow in the polar
waters. These lock together to form a slushy surface layer, known as grease ice,
which hardens and thickens into a continuous sheet. Wind and waves may break
the cover into pieces that jostle and bump to form separate platterlike shapes
of solid ice called pancake ice. These eventually fuse into larger sheets up to
2 m (7 ft) thick, which continually separate into ice floes and then freeze together
again. Near the shore, thick layers of ice that endure for many years may form.

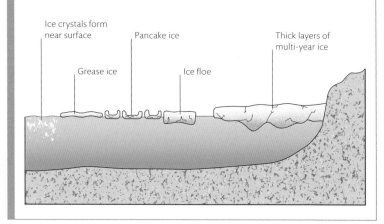

Ice crystals form
near surface

Pancake ice

Thick layers of
multi-year ice

Grease ice

Ice floe

**Swimming legs** are attached to the abdomen

**Large compound eyes** capture light and provide almost 360° vision

**Stomach** is full of phytoplankton (algae)

**Feeding legs** (setae) on the thorax bear fine bristles for sieving food from the water

**Feathery gills** are attached to the thorax

**Primary antennae** are used for touch and smell

**Tips of legs** are used for scraping microscopic algae (diatoms) from the underside of pack ice

**Secondary antennae** are longer and more whiplike than primary antennae

**Underside** of pack ice is often covered with "lawns" of algae on which krill graze

### Seasonal plenty
Antarctic krill (*Euphausia superba*) grow to 5–6 cm (2–2½ in) long on a diet of mostly single-celled algae such as diatoms. They feed by filtering algae from the water using their bristly legs and by raking algae from the underside of pack ice. During winter food shortages, they can reduce body size to limit the chance of starvation.

### Marine biomass
Krill are among the most numerous animals in the ocean, with some swarms of krill containing more than 10,000 individuals per cubic metre (35 cubic feet). Whales, seals, seabirds, fish, and squid consume up to 300 million tonnes of Antarctic krill a year.

# polar swarms

The angle at which Earth tilts as it circles the Sun results in extreme seasonal variation in the amount of light reaching the polar regions. In summer, extended daylight hours create a massive boom in the growth and reproduction of marine phytoplankton (mainly single-celled algae) and consequently also in the animals that feed on them. These include small, shrimplike crustaceans known collectively as krill, which constitute one of the most abundant life-forms and food sources on the planet.

### MARINE FOOD WEB
Phytoplankton live on the surface of the ocean, where they can capture the Sun's energy, which is needed for photosynthesis. As the primary producers, they form the base of the marine food web. In the Antarctic, the phytoplankton are then eaten by krill, which are the primary consumers and hold a critical position in the food web because of the sheer variety of other animals that feed on them. In the Arctic, copepods replace krill as the primary consumers.

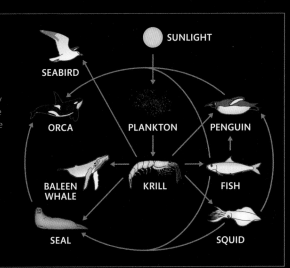

SUNLIGHT

SEABIRD

ORCA · PLANKTON · PENGUIN

BALEEN WHALE · KRILL · FISH

SEAL · SQUID

**Arctic specialist**
The sculpins are a family of small fish that
are common in the icy seas of high northern
latitudes. The antlered sculpin (*Enophrys
diceraus*), native to seas around Alaska and
Siberia, shows some of the highest levels of
blood antifreeze protein known in any fish.

**Eye situated** high on head
enables fish to scan water
above for food and predators

**Upturned
mouth**
adapted for
grabbing food
from below

**Fleshy growths** may
mimic seaweed and
aid camouflage

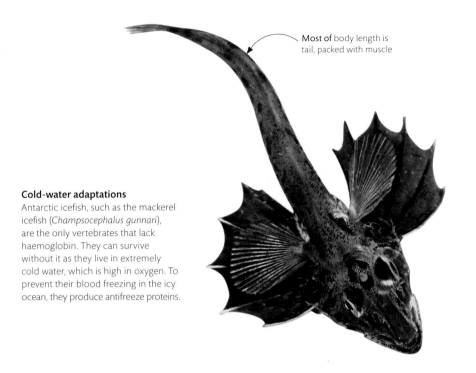

**Most of** body length is
tail, packed with muscle

**Cold-water adaptations**
Antarctic icefish, such as the mackerel
icefish (*Champsocephalus gunnari*),
are the only vertebrates that lack
haemoglobin. They can survive
without it as they live in extremely
cold water, which is high in oxygen. To
prevent their blood freezing in the icy
ocean, they produce antifreeze proteins.

# antifreeze in fish

Since water expands when it turns to ice, freezing can be lethal to living
organisms, causing cells to rupture. Sea water remains liquid to about
-2°C (28°F), and organisms living in sub-zero conditions have evolved
to cope. While mammals and birds burn sugars and fat to keep warm,
many fish produce antifreeze proteins that inhibit ice formation.

**ANTIFREEZE PROTEINS**
Many fish living in polar waters
produce proteins in their blood-
stream that prevent the fish
from freezing. These antifreeze
proteins (AFPs) bind to the
corners of tiny ice crystals
in the blood and prevent the
crystals linking up. As a result,
provided there are enough AFPs
in the fish's body, ice crystals
do not grow sufficiently large
for the blood to freeze.

AFPs bind
to ice crystals

Ice crystals
cannot link
up beyond
a certain size

Crystals free
to link up
to form
solid ice

**ICE CRYSTALS
IN FISH**

**ICE CRYSTALS
IN WATER**

# zooplankton

Animals that drift with ocean currents because they cannot swim or are very weak swimmers are known as zooplankton. They include some of the tiniest animals in the oceans, which, together with phytoplankton (see pp.258–59), form the base of marine food webs. For these creatures, the viscosity of water can make forward progress the equivalent of a human moving through syrup, so they depend on the currents for dispersal. Some are the larvae of animals that are free-swmming as adults, others live out their lives as plankton.

## Fantastical larva
The deepsea-dwelling cusk-eel (*Brotulotaenia nielseni*) lays eggs that float up in the surface, where they hatch. The planktonic larvae live at shallow depths, only making their way to the bottom as they mature. The larva's elaborate fin rays and external gut make it resemble a siphonphore colony with its stinging tentacles, possibly to deter would-be predators.

## Meroplankton
This group of zooplankton are plankton for part of their lives only, usually the larval stage, and many bear little or no resemblance to their final adult form. They feed on other plankton or may live off the yolk of the egg from which they hatched. As adults they may either move to the benthic (deeper ocean) waters or stay in open water.

Tail-like **abdomen** folds underneath the adult crab during metamorphosis

**MEGALOPA STAGE BOX CRAB**
*Calappidae*

**Convoluted bands** of microscopic beating hairs (cilia) aid propulsion

**TORNARIA LARVA**
*Ptychodera flava*

**Transparent mantle** contains pigment-containing cells (chromatophores) that form early in development

**WONDERPUS OCTOPUS LARVA**
*Wunderpus photogenicus*

## Holoplankton
Animals that remain as plankton for their entire life cycle are known as holoplankton. They live in the open (pelagic) water column, and, like meroplankton, most are transparent. They vary in shape and size and are equipped with a variety of adaptations, from body shape to gas-filled floats, to assist survival.

**Spindly, antennae-like** appendages twitch so they appear to jump through water

**BRISTLEWORM**
*Tomopteris helgolandica*

**Fleshy, winglike flaps** (parapodia) help sea slug propel body through water

**SEA ANGEL**
*Clione limacina*

**Flat-shelled snail** that can be up to 3 cm (1 ¼ in) long

**SEA BUTTERFLY**
*Clio recurva*

**Dorsal and anal fins** resemble a bird's feather

**Trailing external gut** increases the surface area for absorption from ingested food

**Large eyes** point upwards, enabling fish to see food falling from above

**JUVENILE DEEPSEA HATCHETFISH**
*Argyropelecus olfersii*

**Larva** has an eye on either side of its head and swims vertically

**LEFTEYE FLOUNDER LARVA**
*Engyprosopon xenandrus*

**Mirrorlike eyes** are used by this crustacean to locate small animal prey

**Full intestine** visible through transparent swimming body of this snail

**Surface** is covered with cartilaginous tubercles

**GLASS SQUID**
*Cranchia scabra*

**OSTRACOD**
*Gigantocypris muelleri*

**SEA ELEPHANT**
*Cardiapoda placenta*

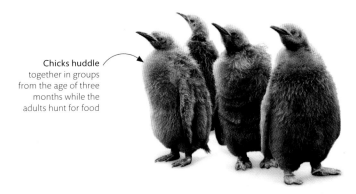

Chicks huddle together in groups from the age of three months while the adults hunt for food

**Contrast in appearance**
Covered in brown fluffy down for the first year of their lives, king penguin chicks bear little resemblance to their parents – so much so that they were once thought to be a completely different species.

*spotlight species*

# king penguin

These birds are found mainly on the subantarctic islands of the southern Atlantic and Indian oceans. There are more than 2.2 million king penguins (*Aptenodytes patagonicus*) around the world, and the largest concentration is on the island of South Georgia.

At up to 1 m (3 ft) tall and weighing up to 16 kg (35 lb), the king penguin is second in size only to its closest relative, the emperor penguin (*A. forsteri*). Its colonies are usually established with close access to the sea, where nests can be built on flat beaches or in tussock grass free of snow and ice.

At 13–16 months, including the adults' pre-moulting period, king penguins have a much longer breeding cycle than any other penguin. They do not build nests. Females lay one egg between November and April, which they pass to their mates, then they return to the sea to hunt. The males carry the eggs (and later, the new chicks) on their feet and cover them with a brood pouch. Males can lose up to 30 per cent of their body weight as they wait for the females to come back. Later, groups of chicks huddle together while both parents go to sea in search of food; feeding sessions are up to 3 months apart, so they too rely on fat reserves to survive. Adults can swim up to 500 km (300 miles) before finding enough fish and squid. There are signs that shifts in the Antarctic Polar Front might increase this distance even further as the penguin's prey moves in response.

A king penguin's eyes are adapted to extreme differences in light conditions. In sunlight, their pupils constrict to a pin-sized opening, but in low light, for example when diving to depths of 300 m (1,000 ft), they expand by 300 times – the widest alteration of pupil size in any bird.

**Group dynamics**
King penguins are highly sociable. Even in the largest colonies, where adults stand flipper to flipper, fighting is rare. However, breeding adults tend to remain separate from non-breeding birds.

### BLUBBER

The blubber of cetaceans and pinnipeds is a thick, relatively firm layer of fatty tissue lying under the skin (made up of the epidermis and dermis). In addition to the adipose (fat) cells, the blubber contains abundant collagen fibres, which stabilize the fat layer, and it is bonded tightly to the underlying muscle by a layer of connective tissue. Blubber thickness ranges from 2 cm (¾ in) in a small seal or a porpoise to about 30 cm (12 in) in large whales.

Dermis · Epidermis

Blubber

Connective tissue

Muscle

**CETACEAN SKIN AND BLUBBER**

### Dispensing with fur

The walrus (*Odobenus rosmarus*) has a thin coat of fur, but by adulthood it often becomes virtually bald, except for a moustache of massive, bristly whiskers. These thick, touch-sensitive hairs are a vital aid to hunting prey, such as clams, in soft sediments.

# insulating layers

For mammals living in cold seas, heat loss can be a constant challenge. Cetaceans (whales, dolphins, and porpoises) and pinnipeds (seals, sea lions, and walruses) rely on a thick layer of fat under the skin, known as blubber, for their survival. The blubber prevents body heat escaping through the skin and also stores energy, including proteins and fats. The thickness and fat content of blubber can be adjusted seasonally, and the blood supply is restricted during dives to minimize heat loss.

### Flexible protection

The skin of a male walrus is anything up to 6 cm (2½ in) thick, overlying a layer of blubber at least as thick again. In addition to insulation, this full body padding provides significant protection and serves as a kind of soft armour when fighting.

**Scars** on skin of males are evidence of "tusking"

Short, sparse hair may
be retained on the head,
even after body hair has
been shed

Nostril linings are the
thinnest part of the skin;
walruses have a well-
developed sense
of smell in air

# southern elephant seal

The world's largest seal, the southern elephant seal (*Mirounga leonina*), is named for its size and the trunklike inflatable proboscis seen on mature males, or bulls. These animals exhibit the greatest gender-related size and weight disparity of all mammals.

Found in cold Antarctic and subantarctic waters, elephant seals feed only at sea and can spend up to ten months a year in their search for fish and squid. They typically dive for 20–30 minutes, although they can can remain underwater for up to 2 hours and reach depths of 2,000 m (6,500 ft) or more. To achieve this, they exhale before diving to eliminate gases that could cause sickness. To ensure a constant oxygen supply, they have twice the amount of oxygen-carrying haemoglobin in their red blood cells as land animals of similar sizes and store more oxygen in their muscles. Their heart rate can slow to 5–15 beats a minute, so blood flows only to vital organs.

Although elephant seals see poorly in daylight, their vision is adapted to hunting in deep, dark water. Their eyes are highly sensitive to the wavelengths of light (see pages 258–59) given off by bioluminescent lanternfish, a main source of food.

Their extended feeding periods ensure these seals accumulate enough fat, or blubber, to see them through two lengthy periods of time on land: a month-long moult between January and February; and the breeding season, which starts around mid-August. The seals return to the same place on land every year. Bulls can spend weeks, or months, fighting for control of harems of females, then breeding with them. The largest bulls acquire vast harems of 50 females or more.

### Coming ashore for breeding

A male southern elephant seal hauls himself out on the island of South Georgia, also the site of a vast king penguin breeding colony. More than 50 per cent of the global population of these seals come to breed here.

### Size difference

Sexual dimorphism – when males and females of a species show marked contrast in appearance – is extreme in these seals. Males can weigh up to 3 tonnes (6,000 lb); females weigh just 600–800 kg (1,300–1,800 lb).

**Huge males** can be up to 10 times larger than females

**Females mate again** a few days after their pups are weaned

# sensitive whiskers

Mammalian hair serves multiple purposes. In seals it provides individually unique patterning, waterproofing, insulation, and vital sensory information. Seals have long, stiff, highly sensitive whiskers (or vibrissae), which grow from hair follicles that detect the slightest vibration and transmit the information along sensory nerves to the brain. This enables seals to navigate in murky water and pick up on prey or predator movement, and helps them to assess the size of breathing and access holes in the ice.

## THE WHISKER SENSORY SYSTEM

Seal whisker follicles are packed with blood vessels and nerve fibres, which process sensory data and relay information about the physical environment to the brain via the trigeminal nerve.

Whisker

Skin's epidermis

Superficial whisker nerve

Sensory nerve endings

Whisker capsule

Deep whisker nerve

Whisker grows from follicle base

**SEAL WHISKER**

Adult fur is greasy and virtually waterproof

**Summer bask**
Weddell seals (*Leptonychotes weddellii*) leave the sea in the short Antarctic summer to give birth, bask, and moult. Unlike the whiskers, which grow throughout the year, the seal's fur is shed and replaced annually.

Coarse, stiff whiskers enable seal to detect prey up to 180 m (590 ft) away

**Whiskers** of pups are thinner and less coarse than those of adults

**Downy lanugo** provides good insulation before the pup develops blubber

**Seal pup**
The Weddell seal is born with whiskers and is covered in lanugo, a fine, dense fur. After four weeks, the lanugo is replaced by the more waterproof adult coat.

**Hind flippers** of true seals give propulsion when swimming and cannot support the body on land

**Front flippers** are used for steering and adjusting speed when swimming; hind flippers provide propulsion

**Fantastic forms**
This immense iceberg was formed in
Disko Bay, western Greenland, where
the Ilulissat Icefjord flows into the sea.
Ten per cent of the ice that fragments from
Greenland's glaciers comes from this glacier.

# ice-shelves
# and icebergs

Ice-sheets and glaciers lock up huge volumes of water over the
landmasses of the polar regions. In coastal areas of Antarctica
and Greenland, ice-sheets and glaciers extend over the sea as vast
ice-shelves that can form towering cliffs up to 50 m (160 ft) high.
Ice-shelves are composed of dense ice that has accumulated over
hundreds of years and are often a vivid blue, because all the tiny
white air bubbles have been squeezed out. Marine algae frozen
into the ice may give parts of ice-shelves a vivid green coloration.
An iceberg is a section of an ice-shelf that has broken away from
the main body of ice. Ice-shelves in both the north and south
polar regions are diminishing in size due to global warming,
releasing massive volumes of water into the world's oceans.

## ICEBERG FORMATION

Ice-shelves and glaciers are grounded on land, while the outer parts float on the
sea. The movement of the floating section with the rise and fall of the tides causes
great strain, especially where crevices are already present. Huge cracks open up,
and chunks of ice break away, which crash into the sea as icebergs – a process
known as calving. Some icebergs can be as large as a small country. More than
50,000 large icebergs are calved from Greenland alone each year.

Snowfall adds to
the thickness of
the ice-sheet

Tidal movement
causes cracking

Floating section
of ice-shelf

Ice-sheet moves
towards the sea

Icebergs break
away

**Sea bear**
The polar bear lives, hunts, and breeds on sea-ice or in water, coming ashore only when forced by melting ice. The fur loses a large proportion of its insulating properties in water, but the dense underfur still provides some benefit by trapping body heat, ensuring the bear stays warm.

**Huge paws** act as shovels, snow shoes, and powerful paddles when swimming

**Arctic giant**
An adult male polar bear on two legs stands over 3 m (10 ft) tall and can weigh over half a tonne. Along with the bear's fur and a layer of blubber, huge size is an adaptation to cope with extreme cold.

# multipurpose fur

Many polar mammals protect themselves from extreme cold by having a coat of fur, but the insulation provided by the fur of different animals varies greatly. The fur of the polar bear is highly effective. It has two layers: a water-repellent outer coat, consisting of long guard hairs, and a dense underfur that keeps the bear warm. The guard hairs are tapered and hollow, making the coat lighter and more buoyant than it would otherwise be. Although polar bears look white, and blend into their Arctic surroundings of snow and ice, their skin is black and their hair is colourless.

## COLOUR AND WARMTH

The guard hairs of the polar bear's coat are transparent. Light falling on them is scattered and reflected so that the bear appears white. Like all fur-coated mammals, a polar bear's metabolism generates heat to keep the body warm and functioning. Most of the insulation is provided by the bear's underfur, where air trapped between the densely packed hairs helps to stop heat escaping from the body on moving currents of air.

Solar radiation

Hollow guard hairs scatter and reflect light

Heat generated in body by bear's metabolism

Most heat escaping from body is trapped in underfur

**POLAR BEAR FUR AND SKIN**

# narwhal

The narwhal (*Monodon monoceros*) can be easily identified by the impressive tusk that gives rise to its nickname the "unicorn of the sea". The tusk is in fact a sensitive, elongated tooth that erupts from the side of the animal's upper jaw.

Narwhals are classed as toothed whales, but the one or, sometimes, two tusks they grow are their only teeth. The tusks are highly flexible, "inside-out" teeth – soft on the outside, hard and dense towards the core – with millions of nerve endings near the surface. Found mainly in males (only around 3 per cent of females have them), they grow in a counterclockwise spiral throughout an animal's life. It was once thought a narwhal used its tusk only for defence, but studies show that it is also a sensory organ. Its nerves detect information from the water, helping the animal to "read" its environment for the presence of food or mates and measure water salinity. A narwhal does not use its tusk to spear food; instead it uses it to tap and stun its prey, which it swallows whole.

Male narwhals become sexually mature around the age of nine years (females mature at six to seven years), when they reach 4–4.5 m (13–15 ft) long and weigh 1,000–1,600 kg (2,200–3,500 lb). The animals live in the Arctic waters of Russia, Norway, Greenland, and Canada. Around 75 per cent of the world's population inhabit the biologically rich waters of Baffin Bay and the Davis Strait, between Canada and Greenland. In winter, when the sea turns to pack ice, they can dive up to 1,100 m (3,600 ft) to hunt for deepwater fish such as halibut, before coming to surface at cracks or holes in the ice to breathe.

### Coming up for air
In winter narwhals congregate, usually in pods of two to 20, in areas of Arctic pack ice. When they surface, like these males at Nunavut, Canada, they are vulnerable to predators such as polar bears.

### THE NARWHAL TUSK
Nerves deep within the core of the narwhal's tooth, or tusk, are connected to tiny pores in its outer surface, or cementum, via an estimated 10 million tubules that lie in the dentine layer. When sea water washes through these tubules, specialized cells at their base detect changes in water composition, pressure, and temperature and relay the information to the brain via the tusk's nerve cells.

Spiral surface on outer layer (cementum) · Hard tissue · Nerves · Veins · Layer of dentine around tubules · Soft tissue · Dentine · Arteries at centre of tusk

**NARWHAL TUSK CROSS SECTION**

# glossary

ABDOMEN The hind part of the body of a vertebrate or arthropod.

ABORAL On the opposite side of the body from the mouth; used especially with reference to animals that have radial symmetry, such as jellyfish and echinoderms. See also *Radial symmetry*.

ABYSSAL A term applied to the depths of the ocean. The abyssal plain is the almost flat plain of the deep ocean floor beyond the continental margin. The abyssal zone is the zone of water between 4,000 and 6,000 m (13,000 and 20,000 ft) deep. See also *Hadal zone*.

ADIPOSE FIN A small, fleshy fin between the dorsal and tail fins in some bony fish.

AERIAL ROOT A root that grows from, or reaches, a point above ground level, as in many mangrove trees. See also *Mangroves*.

ALGA (pl. ALGAE) Any organism that photosynthesizes but is not a true plant, including seaweeds and many microscopic forms. Cyanobacteria are usually now excluded from the definition. See also *Cyanobacteria*.

AMPHIPODS An order of mainly small crustaceans that includes sandhoppers and their marine and freshwater relatives.

AMPULLAE OF LORENZINI The sensory receptors in the skin of sharks and relatives, located around the head region. They can detect weak electric fields such as those produced by the muscle movements of potential prey.

ANADROMOUS Having a life cycle that involves migrating from the sea to fresh water in order to breed, such as in some species of salmon. See also *Catadromous*.

ANAL FIN A single fin on the underside of many fish, near the tail.

ANEMONE See *Sea Anemone*

ANTENNA (pl. ANTENNAE) One of a pair, or in some groups two pairs, of long, thin sensory structures on the heads of arthropods. Antennae are sometimes called "feelers" because of this sensory characteristic.

ARTHROPODS The phylum of jointed-legged invertebrates including crustaceans, insects, and spiders.

ASEXUAL REPRODUCTION Any form of reproduction that does not involve the fusion of sex cells to create a new individual. An example of this is the budding off of new individuals from a parent, which can occur in marine invertebrates, such as some corals.

BACKSHORE The part of the shore lying above the normal high-tide line, reached by the sea only under exceptional conditions.

BACKWASH The flow of water back to the sea after a wave has broken. See also *Swash*.

BALEEN A horny material that grows in the mouths of many whales. It takes the form of plates with hairlike fringes that are used in filter-feeding.

BANK In oceanography, a submerged shallow region or plateau of seafloor surrounded by deeper water.

BARBEL A fleshy, sensitive feeler, or a longer projection, found near the mouth in some fish, such as catfish and dragonfish.

BARNACLES A class of filter-feeding marine crustaceans that live attached to surfaces and have bodies protected by chalky plates.

BARRAGE A low dam with sluices (gates for water) built across a river or estuary. A tidal barrage is designed to trap water at high tide and obtain hydropower when the water is let out.

BEAK The horny, toothless jaws of a bird, turtle, or cephalopod.

BENTHIC Relating to an object or organism living on the seafloor.

BILATERAL SYMMETRY Body symmetry in which there is a left and right side, a head, and a tail. See also *Radial symmetry*.

BINOCULAR VISION A type of vision in which objects are viewed from two eyes simultaneously, which allows perception of depth and distance.

BIOELECTROGENESIS The creation of electric fields by living organisms, including some fish. Weak electric fields can be used as a form of echolocation, while strong ones can be used to "shock" other organisms, either in attack or in defence.

BIOLUMINESCENCE Production of light by living organisms. Sometimes the light is produced by the organism's own cells, and in other cases by light-producing bacteria.

BIOME Any group of large-scale ecosystems across the world that display similar features due to shared climatic conditions. The concept is most commonly applied to land ecosystems.

BIVALVES A class of molluscs that features a pair of hinged shells. Bivalves include clams, mussels, scallops, oysters, and relatives.

BLACK SMOKER A type of hydrothermal vent in which the hot water is coloured dark by sulphide minerals. See also *Hydrothermal vent*.

BLOWHOLE The nostrils of a whale or dolphin, found on the top of the head. The blowhole may be single or double depending on the species.

BLUE CARBON Carbon that is removed from the atmosphere by

ocean plants such as mangroves, seaweeds, and seagrasses.

BONY FISH The class of vertebrates that includes all living fish except hagfish, lampreys, and sharks and their relatives. These fish have an internal skeleton, which in almost all species is made of bone. See also *Cartilaginous fish*.

BREAKWATER A long, solid structure built into the sea to protect a harbour from rough ocean conditions.

CALCAREOUS Containing or made of calcium carbonate.

CALYX In zoology, the cuplike structure of a sea lily or feather star.

CAMOUFLAGE Features of an organism that make it difficult to detect visually against its natural background, including colour patterns and aspects of body shape.

CARAPACE The protective shield covering the upper surface of a turtle, crustacean, or horseshoe crab.

CARRION Animal matter that is dead or decaying.

CARTILAGINOUS FISH A class of fish whose skeleton is made of cartilage, not bone. It includes sharks, rays, and chimaeras.

CATADROMOUS Having a life cycle that involves migrating from fresh water to the sea in order to breed. See also *Anadromous*.

CAUDAL FIN The tail fin of a fish.

CELLULOSE A fibrous carbohydrate that is the main structural substance of plants, many algae, and also tunicates. Animals that consume cellulose through eating plant matter can normally only digest it with the aid of symbiotic bacteria in their gut. See also *Symbiosis, Tunicates*.

CEPHALOPODS A class of tentacled molluscs that includes squid, cuttlefish, octopuses, and nautiluses.

CEPHALOTHORAX The fused head and thorax of a crustacean or a spider.

CHELA A pincer, especially one belonging to an arthropod such as a crab. A chelate leg is a leg with a pincer at the end of it.

CHEMOSYNTHESIS The process by which an organism can grow and multiply using the energy stored in simple chemicals such as hydrogen sulphide or methane. Chemosynthesis functions in contrast to photosynthesis, which depends on energy coming from the Sun. Many bacteria can carry out chemosynthesis, notably those living around hydrothermal vents. See also *Hydrothermal vent, Photosynthesis.*

CHITIN A nitrogen-containing carbohydrate material that forms the skeletal parts of some animals, including the exoskeletons of arthropods. It usually mixes other materials to increase its hardness and strength. See also *Exoskeleton.*

CHLOROPHYLL The green pigment that plants, algae, and some other organisms, including cyanobacteria, use to trap the Sun's energy for use in photosynthesis. There are several different types of chlorophyll. See also *Cyanobacteria, Photosynthesis.*

CHORDATES A phylum of animals that includes all the vertebrates as well as some invertebrate relatives, including tunicates. See also *Tunicates.*

CHROMATOPHORE A pigment-containing cell or group of cells in the skin of many animals. Different types of chromatophores display different colours. Animals with chromatophores can adjust the distribution of pigments within individual cells, altering their overall appearance. See also *Cyanophore, Iridophore.*

CIGUATERA Poisoning from eating toxic sea food, often caused by consuming fish that have eaten toxic dinoflagellates. See also *Dinoflagellates.*

CILIA (sing. CILIUM) Microscopic beating hairs on the surface of some cells, found on animals such as fan worms. These hairs are used for locomotion or to generate currents.

CIRRI (sing. CIRRUS) Small, stiff, hairlike projections in some animals, often used for grasping or movement. They are found, for example, in sea lilies and feather stars.

CLASPERS In sharks and their relatives, the modified pelvic fins of the male that transfer sperm to the female's body during copulation. See also *Pelvic fins.*

CLASS A category in biological classification. It is above an order and below a phylum. See also *Order, Phylum.*

CLOACA A combined body outlet for urinary products, faeces, and reproductive cells, found in most fish and birds.

CNIDARIANS A phylum of invertebrate animals with stinging tentacles that includes sea anemones, corals, jellyfish, and relatives.

COAST See *Emergent coast, Erosional coast, Submergent coast.*

COMPOUND EYE An eye consisting of many small individual facets packed together. Compound eyes are a common type of eye in arthropods.

CONTINENTAL CRUST The part of Earth's crust that forms the continents. It is less dense and thicker than oceanic crust. See also *Oceanic crust.*

CONTINENTAL MARGIN The region of seafloor surrounding a continent, including the continental shelf, continental slope, and continental rise.

CONTINENTAL RISE The region where the deep ocean seabed begins to get shallower at the edge of a continental margin.

CONTINENTAL SHELF The relatively flat and shallow seafloor surrounding a continent, which is geologically part of that continent.

CONTINENTAL SLOPE The sloping seabed that leads from the edge of a continental shelf down to the continental rise.

CONVECTION Circulating movements in a fluid caused by heat being applied to parts of it. This results in warmer and therefore less dense fluid rising upwards, which drives the circulation.

COPEPODS A class of small swimming crustaceans that form an important part of the plankton.

CORALLITE The cup-shaped skeleton of aragonite below the body of an individual coral polyp.

COUNTERSHADING Coloration of an animal's body that helps to disguise it by counteracting the effect of shadows, usually by being lighter on the underside of the body.

CRUSTACEANS A subphylum of arthropods including crabs, lobsters, shrimps, barnacles, and copepods.

CRYPSIS Features or behaviours that allow an animal to avoid detection. Crypsis includes concealed habitats, camouflage, or nocturnal behaviour.

CRYPTIC COLORATION See *Camouflage, Crypsis.*

CTENOPHORES A jellyfishlike phylum of floating carnivorous animals, also called comb jellies.

CURRENT A regular large-scale horizontal or vertical flow of ocean water in one direction; also a smaller flow of water near the shoreline.

CYANOBACTERIA An abundant group of microscopic photosynthesizing bacteria, formerly called blue-green algae. Their tiny cells lack nuclei, as do those of other bacteria. See also *Alga, Protists.*

CYANOPHORE A blue chromatophore. See also *Chromatophore.*

DEPOSIT FEEDER An animal that feeds by extracting food particles from mud or sand.

DETRITIVORE An animal whose food consists of small fragments of dead material (detritus).

DIATOMS A group of microscopic single-celled organisms featuring an ornamented, boxlike protective covering. Most are photosynthesizers and live near the water surface as plankton, where they form an important part of many marine food chains. See also *Photosynthesis, Phytoplankton, Plankton.*

DINOFLAGELLATES A group of microscopic single-celled organisms with two flagella that help them move, many of which are photosynthetic and live near the surface as plankton. See also *Flagellum, Photosynthesis, Plankton.*

DORSAL Relating to the upper side or back of an animal. See also *Ventral.*

DORSAL FIN A fin situated on the back of a fish or whale.

DOWNWELLING The sinking down of water from the surface of the ocean. Large-scale downwelling in certain regions gives rise to a process known as thermohaline circulation. See also *Thermohaline circulation, Upwelling.*

DRIFT A large-scale flow of surface water that is broader and slower-moving than a current.

DUNE A mound of sand formed by the action of winds blowing towards coastal land.

ECHINODERMS A marine invertebrate phylum that includes starfish, sea urchins, sea cucumbers, and relatives. These animals feature tube feet that enable some of them to walk. See also *Tube feet*.

ECHOLOCATION A method of detecting and locating objects, which is used by dolphins and some other animals. Sound waves are emitted and their echoes are then gathered and interpreted. See also *Melon*.

EMERGENT COAST A coast where the land has risen relative to the current sea level. See also *Submergent coast*.

ENDOPARASITE A parasite that lives inside the body of another organism. See also *Parasite*.

ENDOSKELETON A skeleton that is inside the body rather than on the surface, such as the skeletons of vertebrates and echinoderms.

EPITHELIUM A layer of living cells on the surface of an animal's body or of an internal organ.

EROSIONAL COAST A coastline that is gradually being eroded by the sea.

EURYHALINE Able to survive in water of a wide range of salinities. See also *Salinity, Stenohaline*.

EXOSKELETON A skeleton on the outside of the body, such as the skeletons of arthropods.

EXTERNAL FERTILIZATION Fertilization in which egg and sperm combine outside the body, usually in sea water. See also *Internal fertilization*.

FAMILY A category in biological classification. It is above a genus and below an order. See also *Genus, Order*.

FAST ICE Sea-ice that is frozen on to the coast or land-formed ice at its landward edge. See also *Sea-ice*.

FILTER-FEEDER An animal that feeds by filtering small food particles from its environment, through the baleen in the case of some whales. Filter-feeders are a type of suspension-feeder. See also *Baleen, Suspension-feeder*.

FJORD An inlet of the sea once occupied by a glacier. Fjords are often deep and steep-sided, with a shallower sill at their mouth. See also *Glacier*.

FLAGELLUM (pl. FLAGELLA) A long, microscopic whiplike structure used by some single-celled organisms, such as dinoflagellates, for locomotion. See also *Dinoflagellates*.

FLIPPER A limb used for swimming, for example in whales, seals, penguins, and turtles.

FLUKE Either of the two flattened sides of a whale's tail.

FORAMINIFERANS Single-celled organisms that mainly live on the seafloor, protected by chalky shells, and feed on other small organisms.

FRAZIL ICE Ice that forms in cold water and consists of thousands of tiny crystals. It is an early stage in the development of sea-ice and in turbulent waters it can form pancake ice. See also *Pancake ice, Sea-ice*.

FROND A flattened, leaflike structure found in many seaweeds. Unlike the leaves of land plants, seaweed fronds do not contain specialized transport vessels to interchange materials with the rest of the organism.

GASTROPODS The largest class of molluscs, which includes all snails and slugs.

GENUS (pl. GENERA) A category in biological classification. It is above

a species and below a family. The scientific name of a species consists of two words: a genus name and a species epithet. For example, *Ursus maritimus*, the polar bear, is a species in the genus *Ursus*, which also includes other bears. See also *Family, Species*.

GILL A structure for absorbing oxygen from water, often found in fish, molluscs, and crustaceans. Gills can also have other functions such as sieving small food particles.

GLACIER A "river" of ice that flows slowly downhill from an ice cap or mountainous region.

GUYOT A sea mount with a flat top. See also *Sea mount*.

GYRE A large system of ocean currents circulating round a centre point. There are five major ocean gyres.

HADAL ZONE The deepest zone of ocean water, below 6,000 m (20,000 ft), which occurs only in ocean trenches. See also *Abyssal*.

HERMAPHRODITE An animal that is both male and female. A simultaneous hermaphrodite is male and female at the same time, while a sequential hermaphrodite changes from one sex to the other, either once or repeatedly.

HOLDFAST The structure that holds a seaweed to the seafloor. It is not a true root because it does not absorb substances for use in the rest of the body.

HOST Any organism on which a given parasite regularly feeds.

HYDROIDS A subgroup of hydrozoans that usually grow as small, plantlike colonies of connected individuals. See also *Hydrozoans*.

HYDROTHERMAL VENT A fissure in a volcanically active area of the ocean floor, where water rich in

chemicals issues from the rock at high temperatures.

HYDROZOANS A class of mainly colonial cnidarians that includes hydroids, siphonophores, and some small jellyfish. See also *Cnidarians, Hydroids, Jellyfish, Siphonophores*.

ICE-SHEET A very large area of ice that permanently covers a land area, as in Antarctica or Greenland.

ICE-SHELF A floating extension of an ice-sheet or glacier over the ocean.

INTERNAL FERTILIZATION Fertilization where the eggs and the sperm meet and fuse inside the body of an animal. See also *External fertilization*.

INTERTIDAL Between the high and low water marks of a shoreline.

INVERTEBRATE Any animal other than a vertebrate without a backbone, such as tunicates and arthropods. See also *Vertebrates*.

IRIDOPHORE A reflective or iridescent kind of chromatophore. See also *Chromatophore*.

ISOPODS An order of mainly small crustaceans that includes woodlice, sea slaters, and many marine species.

JELLYFISH Members of the cnidarian class Scyphozoa, whose adults drift or swim slowly in the ocean and use stinging cells to subdue prey. The deadly "box jellyfish" are a separate but related class, and the word can also be used for the small medusas of hydrozoans. Despite their name jellyfish are not, in fact, fish. See also *Cnidarians, Hydrozoans, Medusa*.

KERATIN A tough protein that forms the substance of claws, nails, and hair.

KINGDOM The highest category in traditional biological classification.